Stanislav Bulygin

Computer algebra in coding theory and cryptanalysis

Stanislav Bulygin

Computer algebra in coding theory and cryptanalysis

Polynomial system solving for decoding linear codes and algebraic cryptanalysis

Südwestdeutscher Verlag für Hochschulschriften

Impressum / Imprint
Bibliografische Information der Deutschen Nationalbibliothek: Die Deutsche Nationalbibliothek verzeichnet diese Publikation in der Deutschen Nationalbibliografie; detaillierte bibliografische Daten sind im Internet über http://dnb.d-nb.de abrufbar.
Alle in diesem Buch genannten Marken und Produktnamen unterliegen warenzeichen-, marken- oder patentrechtlichem Schutz bzw. sind Warenzeichen oder eingetragene Warenzeichen der jeweiligen Inhaber. Die Wiedergabe von Marken, Produktnamen, Gebrauchsnamen, Handelsnamen, Warenbezeichnungen u.s.w. in diesem Werk berechtigt auch ohne besondere Kennzeichnung nicht zu der Annahme, dass solche Namen im Sinne der Warenzeichen- und Markenschutzgesetzgebung als frei zu betrachten wären und daher von jedermann benutzt werden dürften.

Bibliographic information published by the Deutsche Nationalbibliothek: The Deutsche Nationalbibliothek lists this publication in the Deutsche Nationalbibliografie; detailed bibliographic data are available in the Internet at http://dnb.d-nb.de.
Any brand names and product names mentioned in this book are subject to trademark, brand or patent protection and are trademarks or registered trademarks of their respective holders. The use of brand names, product names, common names, trade names, product descriptions etc. even without a particular marking in this work is in no way to be construed to mean that such names may be regarded as unrestricted in respect of trademark and brand protection legislation and could thus be used by anyone.

Verlag / Publisher:
Südwestdeutscher Verlag für Hochschulschriften
ist ein Imprint der / is a trademark of
OmniScriptum GmbH & Co. KG
Heinrich-Böcking-Str. 6-8, 66121 Saarbrücken, Deutschland / Germany
Email: info@svh-verlag.de

Herstellung: siehe letzte Seite /
Printed at: see last page
ISBN: 978-3-8381-0948-0

Zugl. / Approved by: Kaiserslautern, TU, Diss., 2009

Copyright © 2009 OmniScriptum GmbH & Co. KG
Alle Rechte vorbehalten. / All rights reserved. Saarbrücken 2009

Overview

This book that represents the author's Ph.D. thesis is devoted to applying symbolic methods to the problems of decoding linear codes and of algebraic cryptanalysis. The paradigm we employ here is as follows. We reformulate the initial problem in terms of systems of polynomial equations over a finite field. The solution(s) of such systems should yield a way to solve the initial problem. Our main tools for handling polynomials and polynomial systems in such a paradigm is the technique of Gröbner bases and normal form reductions.

The first part of the book is devoted to formulating and solving specific polynomial systems that reduce the problem of decoding linear codes to the problem of polynomial system solving. We analyze the existing methods (mainly for the cyclic codes) and propose an original method for arbitrary linear codes that in some sense generalizes the Newton identities method widely known for cyclic codes. We investigate the structure of the underlying ideals and show how one can solve the decoding problem – both the so-called bounded decoding and more general nearest codeword decoding – by finding reduced Gröbner bases of these ideals. The main feature of the method is that unlike usual methods based on Gröbner bases for "finite field" situations, we do not add the so-called field equations. This tremendously simplifies the underlying ideals, thus making feasible working with quite large parameters of codes. Further we address complexity issues, by giving some insight to the Macaulay matrix of the underlying systems. By making a series of assumptions we are able to provide an upper bound for the complexity coefficient of our method. We address also finding the minimum distance and the weight distribution. We provide solid experimental material and comparisons with some of the existing methods in this area.

In the second part we deal with the algebraic cryptanalysis of block iterative ciphers. Namely, we analyze the small-scale variants of the Advanced Encryption Standard (AES), which is a widely used modern block cipher. Here a cryptanalyst composes the polynomial systems which solutions should yield a secret key used by communicating parties in a symmetric cryptosystem. We analyze the systems formulated by researchers for the algebraic cryptanalysis, and identify the problem that conventional systems have many auxiliary variables that are not actually needed for the key recovery. Moreover, having many such auxiliary variables, specific to a given plaintext/ciphertext pair, complicates the use of several pairs which is common in cryptanalysis. We thus provide a new system where the auxiliary variables are eliminated via normal form reductions. The resulting system in key-variables only is then solved. We present experimental evidence that such an approach is quite good for small scaled ciphers. We investigate further our approach and employ the so-called meet-in-the-middle principle to see how far one can go in analyzing just 2–3 rounds of scaled ciphers. Additional "tuning techniques" are discussed together with experimental material. Overall, we believe that the material of this part of the thesis makes a step further in algebraic cryptanalysis of block ciphers.

Preface

In this book we consider the decoding problem of linear codes and cryptanalysis of small scale variants of Advanced Encryption Standard (AES), a modern and widely used block cipher. There are a vast variety of methods employed in both fields; we concentrate on symbolic methods, namely methods based on solving systems of polynomial equations over a finite field.

The book is organized as follows. Chapter 1 is devoted to questions of polynomial system solving for decoding. We start with giving necessary background on codes and Gröbner bases in Sections 1.1 and 1.2. Further we give an overview of some of the existing methods for decoding (mainly cyclic) codes in Section 1.3. This material appears also in [34]. Then follows the main part of the chapter: Section 1.4. Here we present a new method of decoding arbitrary linear codes via solving a system of quadratic equations. Some technical preliminaries are given in Sections 1.4.1 and 1.4.2. The main results that apply to decoding are presented in Section 1.4.3. The results presented here appear in [31, 32, 33]. Our method in some sense is a generalization of Newton identities method illustrated in Section 1.3.2; this is made explicit in Section 1.4.4. The method of quadratic equations can be adjusted to find the minimum distance of a linear code: Section 1.4.5 is devoted to this question. We briefly discuss a possibility of generic decoding for linear codes in Section 1.4.6. Usually estimating complexity of methods based on non-linear polynomial system solving is a very challenging task. We try to shed some light on this question in Section 1.4.7. We further provide a solid experimental material on our method in Section 1.4.8. We end the chapter with conclusions and possible future work directions. The new contribution in Chapter 1 is in Section 1.4. Here the initial idea belongs to Ruud Pellikaan, however the further development of ideas is hard to be assigned to anyone specifically, except that Section 1.4.2 is work out primarily by Ruud Pellikaan and Sections 1.4.6, 1.4.7 and 1.4.8 primarily by the author. Equivalent representation of the quadratic system in Section 1.4.3 is worked out together with Sergiy Ovsyenko during his visits at the University of Kaiserslautern in 2006 and 2007.

Chapter 2 is devoted to algebraic cryptanalysis of small scale variants of the AES. After giving an introduction in Section 2.1, we present some background material on block ciphers and algebraic cryptanalysis in Section 2.2 and on algebraic properties of the AES in Section 2.3. The original method of key-variables only equations is presented in Section 2.4. We do an analysis of the existing polynomial systems for cryptanalysis in Section 2.4.1 and then propose another system in Section 2.4.2; we take a look at its properties and experimental data. We move further to the meet-in-the-middle scenario in Section 2.4.3. Here, if we limit ourselves to 2-3 rounds, we can consider pretty large scaled AES-ciphers. Some refinements together with further experimental data are given in Section 2.4.4. At the end of the chapter we give conclusions and future research directions in this area that we consider as promising. The new contribution in Chapter 2 is in Section 2.4, which appears in [23, 35]. Here the initial ideas were developed together with Michael Brickenstein and the exact distinguishing is hard to make. Further, Section 2.4.3 is worked out primarily by the author and Section 2.4.4 is worked out primarily by Michael Brickenstein.

Acknowledgments

Before starting with the results of my thesis I would like to express my acknowledgment to people who made possible my present work, supported, and inspired me through the whole time devoted to the thesis.

First of all, I would like to thank my supervisor Prof. Gert-Martin Greuel for encouragement and valuable tips for professional life. He is the one who gave me a great opportunity to do my Ph.D. research in a supporting and comforting atmosphere. The list of attended conferences would not be that large if not the generous financial support provided by Prof. Greuel. So I would like to thank him for trusting in me and believing that he is not wasting money on my numerous scientific journeys. I would also like to thank my second supervisor Prof. Gerhard Pfister. He is always a person who can try to (and usually does) answer mathematical questions that arose in the course of the work. I also had a pleasant time to be a teaching assistant for his lectures on cryptography and coding theory as well as elliptic curve cryptography. This duty made me learn certain useful things. Special thanks go to my bachelor thesis supervisor Sergiy Ovsyenko from Kyiv Shevchenko University. He taught me many things in algebra and encouraged me to do coding theory in the first place. His visits to Kaiserslautern helped to clarify certain points in the thesis. For this I am especially grateful.

Next, I would like to say "Thank you so much!" to my coding-mentor Ruud Pellikaan from the Technical University of Eindhoven. It may be said with absolute certainty that the results of this thesis would not appear without my cooperation with this person. I have learned so much during these years and I believe that there is so much to learn left, as Ruud is an inexhaustible source of knowledge in this field. My deep appreciation goes to Dr. Pellikaan. I remember with pleasure working with Michael Brickenstein on the cryptographic questions, which are covered in the second part of my thesis. Michael has a subtle understanding of things. Although it was not always easy to comprehend exactly what he means, after thinking everything through I got excited with some of his ideas.

Here I would like to mention my colleagues and other people who aspired my work. Among them are Edgar Martinez-Moro, Antinio Campillo, Massimiliano Sala, Maria Bras-Amoros, Diego Ruano, Fernando Hernando, Olav Geil, Ludovic Perret, Martin Albrecht, Viktor Levandovskyy, Tapan Rai, and many others. These people made my mathematical view broader and deepened my knowledge of many things.

I should not forget my fellow students who made my stay in Kaiserslautern fun. Thanks for that go to Oleksandr Manzyuk, Oleksandr Yena, then Oleksandr Motsak and Oliver Schmidt (it is a pleasure to share the office with you, guys), Stefan Steidel, Christian Eder, Frank Seelisch. Special thanks go to my friend Dr. Evgeniy Ivanov for making my stay here an unforgettable experience.

Using this chance I would like to thank people at the Group "Algebra, Geometry, and Computer Algebra" for teaching me useful subjects during my master's studies. Among them are my supervisors, and also Dr. Thomas Markwig, Dr. Christoph Lossen, Prof. Ulrich Dempwolff, Dr. Anne Frühbis-Krüger. Thanks to Dr. Hans Schönemann for his patient explanations of some tricky situations around SINGULAR. It seems that there are many questions in this area that only Hans can answer. I would like to mention the secretary of our group Petra Bäsell. This kind woman and a great person is the one always willing to help. Things could be much more difficult without her.

Finally, I would like to show my appreciation to my family: mother Valintina, father Valeriy, sister Vladislava, and grandmother Claudia. They always believe and support me. Their kind and insightful advices are absolutely necessary to go through life with dignity. Thank you so much for this! My very very thanks go to my precious wife Marina. She is always by my side, inspiring me in everything I do. Her presence definitely gives me the "time of my life". My tender and sincere gratitude goes to her.

Post factum: The thesis that is presented in this book was defended at the Department of Mathematics of the University of Kaiserslautern on June, 12, 2009. The author would like to thank all the people who were present there. Special thanks go to Ruud and Marion Pellikaan that honored me by attending the event.

Contents

1 **System solving in decoding**   3
    1.1 Codes and decoding problem 3
    1.2 Gröbner bases for system solving 10
    1.3 Gröbner bases in coding theory: overview 14
        1.3.1 Cooper's philosophy and its development 15
        1.3.2 Generalized Newton identities 22
        1.3.3 Decoding affine variety codes 26
        1.3.4 Syndrome decoding with Gröbner bases 29
    1.4 Methods based on quadratic equations 33
        1.4.1 Matrix in MDS form 34
        1.4.2 Determinantal variety of syndromes 38
        1.4.3 Nearest codeword decoding 46
        1.4.4 Generalized Newton identities for arbitrary linear codes 57
        1.4.5 Finding the minimum distance 59
        1.4.6 Generic decoding 62
        1.4.7 Complexity issues 63
        1.4.8 Simulations and experimental results 78

2 **System solving in cryptanalysis**   89
    2.1 Introduction 89
    2.2 Block ciphers and algebraic cryptanalysis 90
    2.3 AES: Advanced Encryption Standard 94
    2.4 Attacking with key variables 96
        2.4.1 Analyzing the polynomial system for cryptanalysis . . 96
        2.4.2 Gröbner basis shape. Normal forms 100
        2.4.3 Meet-in-the-middle attack 103
        2.4.4 Further optimizations 107

Bibliography   111

Index   121

Notation

C_I	code C punctured at positions from I
$Compl(C)$	complexity coefficient of a decoding algorithm for a code C
$\det(M)$	determinant of a matrix M
$d(C)$	minimum distance of a code C
$d(\mathbf{r}, C)$	distance from a vector $\mathbf{r}$ to a code C
$d(\mathbf{x}, \mathbf{y})$	Hamming distance between vectors $\mathbf{x}$ and $\mathbf{y}$
$\overline{\mathbb{F}}$	algebraic closure of a field $\mathbb{F}$
$\mathbb{F}[X_1, \ldots, X_n]$	polynomial ring over $\mathbb{F}$ with variables $X_1, \ldots, X_n$
$\mathbb{F}_q$	finite field with q elements
$H_q(\cdot)$	q-ary entropy function
$I(\mathbf{k})$	defining ideal of $L(\mathbf{k})$
$I(t, \mathcal{U}, V)$	quadratic part of our system, Definition 1.4.29
$I(t, \mathcal{V})$	ideal generated by the determinants as per Definition 1.4.20
$J(\mathbf{r})$	linear part of our system, Definition 1.4.37
$J(t, \mathbf{r})$	our system, Definition 1.4.37
$J_q(t)$	our system for the generic decoding, Section 1.4.6
$J_q(t, \mathbf{r})$	system $J(t, \mathbf{r})$ with the field equations added
$\mathrm{lc}(f)$	leading coefficient of a polynomial f
$L_C(X_1, \ldots, X_{n-k}, Z)$	general error-locator polynomial of a code C
$L(I)$	leading ideal of an ideal I
$\mathrm{lm}(f)$	leading monomial of a polynomial f
$\mathrm{lt}(f)$	leading term of a polynomial f
$L(\mathbf{k})$	linear space defined by positions from $\mathbf{k}$
$\mathcal{L}(\mathbf{r}, C)$	list of nearest to $\mathbf{r}$ codewords from C
M_v	submatrix of a matrix M consisting of the first v columns
M_{uv}	$u \times v$ submatrix of M given by the right upper corner of M
μ_l^{ij}	structure constant
$\mathrm{NF}(f, G)$	normal form of a polynomial f w.r.t a Gröbner basis G
$\mathrm{rad}(I)$	radical of an ideal I
$\mathrm{rank}(M)$	rank of a matrix M
$SBOX(\cdot)$	implicit S-Box equations
$sbox(\cdot)$	explicit S-Box equations
$\mathrm{spoly}(f, g)$	s-polynomial of polynomials f and g
$\mathbf{s}(\mathbf{r})$	(known) syndrome of a vector $\mathbf{r}$
$SR(n, r, c, e)$	scaled version of AES: n rounds, e-bit words, dimensions $r \times$
$\mathrm{supp}(f)$	support of a polynomial f
$\sigma(Z)$	error-locator polynomial
U_{ij}	linear function as per Definition 1.4.15
$\mathbf{u}(B, \mathbf{e})$	unknown syndrome of a vector $\mathbf{e}$ w.r.t a matrix B
$\mathbf{u}(\mathbf{e})$	unknown syndrome of a vector $\mathbf{e}$
$\mathcal{U}(\mathbf{e})$	matrix of unknown syndromes of $\mathbf{e}$
$V(F)$	variety of a set of polynomials F
$V(t, \mathcal{U}, V)$	variety of $I(t, \mathcal{U}, V)$
$V(t, \mathcal{V})$	variety of $I(t, \mathcal{V})$
$\mathrm{wt}(\mathbf{x})$	Hamming weight of a vector $\mathbf{x}$
$\mathbf{x} * \mathbf{y}$	coordinatewise product of vectors $\mathbf{x}$ and $\mathbf{y}$

Chapter 1

Polynomial system solving for decoding linear codes

1.1 Codes and decoding problem

The problem of robust information transmission is one of the problems that is encountered in modern technological solutions. Due to appearance of noise of different sort, information transmitted can be distorted, thus disabling a receiver to correctly process and utilize it. Some examples of such a situation:

- Satellite and mobile communication, where a signal can be spoiled by various interference in atmosphere or outer space.

- Compact Discs (CDs) and Digital Video Discs (DVDs), where information written on a disc can be affected by mechanical means such as scratches and fingerprints.

- Measuring of electromagnetic data of biological nature, where a signal can be corrupt due to imprecise measuring devices or due to subjective factors.

We would like to be able to recover some portion of damaged information giving up some efficiency. It is possible to arrange that by adding redundant information in a special way, which at a receiver's side gives an opportunity to correct some portion of a message sent.

Linear codes

There are several methods of achieving the idea described above. In this thesis we address the idea of linear block codes over a finite field. Here we present necessary basic background. For more thorough treatment of the subject one is referred to [93, 88].

First we need to fix an alphabet we are going to work with, i.e. messages are to be composed of the elements of this alphabet. So let Q be an alphabet set. One can require different properties from Q depending on what kind of codes one wants to construct. For the case of linear codes that is treated here, we require Q to be a finite field with q elements denoted by $\mathbb{F}_q$. The idea of a block code in general is to divide message to be transmitted into blocks of equal length k and add some redundant information to each block in a systematic way in order to obtain new blocks of length n. These blocks are transmitted and decoded by receiver blockwise to get the initial blocks.

Definition 1.1.1 Let $\mathbb{F}_q$ be a field with q elements. A *linear code* C of *length* n and *dimension* k is a k-dimensional vector subspace of $\mathbb{F}_q^n$. We say that we are working with a q-ary $[n, k]$ code. The elements of C are called *codewords*. The ratio $R := k/n$ is called the *information rate*, whereas $n-k$ is the *redundancy* of a linear code.

The idea now is to map somehow the vector space $\mathbb{F}_q^k$, which represents "plain", unencoded messages to some $[n, k]$ code C ($n \geq k$). This process is referred to as *encoding*. Now the codewords of C are transmitted and affected by noise during the transmission, so that some transmitted elements from C may change (in practice we expect only quite moderate change). The receiver then has to decide, to which codeword of C does the received vector $\mathbf{r}$ corresponds to. This process is referred to as *decoding*. The notion of information rate is well motivated, since R shows what is the fraction of actual information in the encoded message, whereas $1-R$ indicates the fraction of redundancy added. Now let us see how one can define a linear code and how one can encode messages.

Definition 1.1.2 Let C be a q-ary $[n, k]$ code and let $B = \{\mathbf{b}_1, \ldots, \mathbf{b}_k\}$ be any basis of C as a vector subspace of $\mathbb{F}_q^n$. A $k \times n$ matrix G which rows are all vectors from B is called the *generator matrix* of the code C.

Having some generator matrix for the code C we can map $\mathbb{F}_q^k$ to C via the linear map defined by G: $\mathbb{F}_q^k \ni \mathbf{a} \mapsto \mathbf{a}G \in C \subseteq \mathbb{F}_q^n$, where $\mathbf{a}$ is seen as a row vector. We say that two codes C_1 and C_2 are *equivalent* if there exists a permutation $\pi \in S_n$ and a vector $\mathbf{p} = (p_1, \ldots, p_n)$ with non-zero entries such that any codeword $\mathbf{c}_1 \in C_1$ is obtained from some codeword $\mathbf{c}_2 \in C_2$ by multiplying it coordinatewise by the elements of $\mathbf{p}$ and applying π to the coordinate positions. A generator matrix G is said to be in *row reduced echelon form* if it is of the form $G = (I_k | A)$, where I_k is the $k \times k$ identity

1.1. CODES AND DECODING PROBLEM

matrix and A some matrix with entries from $\mathbb{F}_q$. It can be shown that every code has an equivalent code with the generator matrix in row reduced echelon form. Note that if $G = (I_k|A)$ and $H = (-A^T|I_{n-k})$, then $GH^T = 0$. This implies that for every codeword $\mathbf{c} \in C : H\mathbf{c}^T = 0$, moreover this is the defining property for the codewords of C.

Definition 1.1.3 Let C be a q-ary $[n,k]$ code with the generator matrix G. Let H be a full-rank $(n-k) \times n$ matrix with the property $GH^T = 0$, then H is called the *parity check matrix* for the code C.

Definition 1.1.4 Define the inner product of two vectors $\mathbf{x} = (x_1, \ldots, x_n)$ and $\mathbf{y} = (y_1, \ldots, y_n)$ from $\mathbb{F}_q^n$ as $\langle \mathbf{x}, \mathbf{y} \rangle := \sum_{i=1}^n x_i y_i$. Let C be a q-ary $[n,k]$ code. Define $C^\perp := \{\mathbf{x} \in \mathbb{F}_q^n | \langle \mathbf{x}, \mathbf{c} \rangle = 0 \ \forall \mathbf{c} \in C\}$. The code $C^\perp$ is the *orthogonal code* of C.

Obviously, $C^\perp$ is a linear subspace of dimension $n - k$, so the above notion is well-defined. Let the code C have G and H as its generator and parity check matrix resp. From the equality $GH^T = 0$ it follows that H and G are in fact generator and parity check matrices of $C^\perp$ resp.

So far we stayed on the side of the sender and encoder. In order to provide decoding capabilities for our codes we need to enhance the vector space $\mathbb{F}_q^n$ with a special metric, which would reflect how strongly a received vector differs from its sent counterpart.

Definition 1.1.5 The *Hamming metric* $d(\cdot, \cdot)$ on $\mathbb{F}_q^n$ is defined as follows: for two vectors $\mathbf{x} = (x_1, \ldots, x_n)$ and $\mathbf{y} = (y_1, \ldots, y_n)$ from $\mathbb{F}_q^n$ we have $d(\mathbf{x}, \mathbf{y}) := |\{i | x_i \neq y_i\}|$. The *Hamming weight* of a vector $\mathbf{x} \in \mathbb{F}_q^n$ is defined as $\text{wt}(\mathbf{x}) := |\{i | x_i \neq 0\}|$. For a q-ary $[n,k]$ code C we define its *minimum distance* as $d = d(C) := \min_{\mathbf{x}, \mathbf{y} \in C, \mathbf{x} \neq \mathbf{y}} d(\mathbf{x}, \mathbf{y})$. The notation for the code C now is $[n, k, d]$. The ratio $\delta = d/n$ is called *relative minimum distance* of C. The *minimum weight* of C is defined as $\text{wt}(C) := \min_{\mathbf{x} \in C, \mathbf{x} \neq 0} \text{wt}(\mathbf{x})$.

It can be shown that the function $d(\cdot, \cdot)$ so defined is indeed a metric on $\mathbb{F}_q^n$. Linear codes possess the following nice property which is easy to check: for every linear code $C : d(C) = \text{wt}(C)$.

Now, suppose that we have chosen the q-ary $[n, k, d]$ for encoding. We have that the Hamming distance between any two codewords is at least d. Let us imagine that we have constructed balls in Hamming metric (which historically are called spheres and *vice versa*) with centers in the codewords of C and radius $e := \lfloor (d-1)/2 \rfloor$. These balls do not intersect. Now, assume that during the transmission of a codeword $\mathbf{c}$ at most e errors occurred, i.e. the received vector $\mathbf{r}$ differs from $\mathbf{c}$ in at most e positions, or, equivalently, is at

Hamming distance at most e to **c**. This means that **r** lies in the ball described above with center in **c**. If the decoder observing **r** has to make a decision on which codewords has actually been sent, it is plausible to take the closest codeword w.r.t Hamming distance. In our case this would be the codeword **c**. This decoding principle is called *maximum-likelihood-decoding*. Note that since the balls do not intersect, the decoder always makes a unique correct decision under assumption that at most e errors occurred. The parameter e is called *error-correcting capacity* of the code. The following is an example of one-error correcting code due to Hamming.

Example 1.1.6 Consider the matrix over $\mathbb{F}_2$:

$$H = \begin{pmatrix} 0 & 0 & 0 & 1 & 1 & 1 & 1 \\ 0 & 1 & 1 & 0 & 0 & 1 & 1 \\ 1 & 0 & 1 & 0 & 1 & 0 & 1 \end{pmatrix},$$

which columns are all non-zero column vectors from $\mathbb{F}_2^3$. Define a code C such that H is its parity check matrix. Since H is full-rank, we have that C is a binary $[7,4]$ code. In order to find its minimum distance observe that any two columns of H are linearly independent, thus $d(C) > 2$, otherwise we would get a contradiction with $\mathbf{c} \in C \iff H\mathbf{c}^T = 0$. On the other hand, e.g. $\mathbf{c} = (1001100)$ is a codeword of weight 3, so $d(C) = 3$. Therefore, C has parameters $[7, 4, 3]$ and thus can correct $(3-1)/2 = 1$ error. It can be shown that the balls of radius one with centers in codewords of C cover the entire $\mathbb{F}_2^7$. The codes with such property are called *perfect*.

The following result shows [50] that for codes of large length it is possible to compute the parameters almost always.

Theorem 1.1.7 *Let R be such that $0 < R < 1$. The fraction of q-ary $[n, \lfloor nR \rfloor]$ linear codes that have relative minimum distance*

$$\delta = H_q^{-1}(1-R),$$

where $H_q(x) = -x \log_q x - (1-x) \log_q(1-x) + x \log_q(q-1)$ is the q-ary entropy function (we denote $H(x) := H_2(x)$), tends to one, as n tends to infinity.

Cyclic codes

Now we give a very brief overview of an important class of linear codes that are called cyclic codes [93, 88, 108].

1.1. CODES AND DECODING PROBLEM

Definition 1.1.8 The linear code C is *cyclic*, if for every codeword $\mathbf{c} = (c_0, \ldots, c_{n-1})$ in C its cyclic shift $(c_{n-1}, c_0, \ldots, c_{n-2})$ is again a codeword in C.

When working with cyclic codes, vectors are usually presented as polynomials. So $\mathbf{c}$ is represented by the polynomial $c(X) = \sum_{i=0}^{n-1} c_i X^i$ with $X^n = 1$. Denote by x an image of X in the factor ring $\mathbb{F}_q[X]/\langle X^n - 1 \rangle$ under natural homomorphism. So $c(x)$ is an element of the factor ring $\mathbb{F}_q[X]/\langle X^n - 1 \rangle$. Cyclic codes over $\mathbb{F}_q$ of length n correspond one-to-one to ideals in this factor ring. A multiplication of a codeword polynomial by X corresponds to a cyclic shift by one position to the right. We assume for cyclic codes that $(q, n) = 1$. Let $\mathbb{F} = \mathbb{F}_{q^m}$ be the splitting field of $X^n - 1$ over $\mathbb{F}_q$. Then $\mathbb{F}$ has a *primitive n-th root of unity* which will be denoted by a. Since $(q, n) = 1$, the polynomial $X^n - 1$ splits in $\mathbb{F}$ completely with different linear factors. A cyclic code is uniquely given by a *defining set* S_C which is a subset of $\mathbb{Z}_n$ such that

$$c(x) \in C \text{ if } c(a^i) = 0 \text{ for all } i \in S_C.$$

A code has several defining sets. The *complete defining set* of C is the set of all $i \in \mathbb{Z}_n$ such that $c(a^i) = 0$ for all $c(x) \in C$. If $c(a^i) = 0$, then $c(a^{qi}) = (c(a^i))^q = 0$. Hence a defining set is complete if and only it is invariant under multiplication by q. A *cyclotomic set* of a number $j \in \mathbb{Z}_n$ is a subset $Cl(j) := \{jq^i \mod n | i \subset \mathbb{N}\}$. A defining set is complete if and only if it is a disjoint union of some cyclotomic sets. The size of the complete defining set is equal to the redundancy $r = n - k$. The following example show that the Hamming code from Example 1.1.6 actually can be seen as a cyclic code.

Example 1.1.9 Let a be a primitive 7th root of unity in $\mathbb{F}_8$. We want to show that the cyclic code $C' := \{c(x) | c(a) = 0\}$ is equivalent to the code C from Example 1.1.6. Indeed, Since $a^i \notin \mathbb{F}_2$ for all $i = 1, \ldots, 6$, it follows that the columns of the matrix H', which are the representations of $1, a, a^2, \ldots, a^6$ as vectors in $\mathbb{F}_2^3$, are pairwise linearly independent over $\mathbb{F}_2$. The condition $c(a) = 0$ for $c \in C'$ is equivalent with $H' c^T = 0$, where c is seen as a row vector. So H' is the parity check matrix of the code equivalent to C. The complete defining set of C' is $\{1, 2, 4\}$.

Decoding problem

We have already discussed the decoding problem, when we assumed that the number of errors occurred is at most $e = \lfloor (d(C) - 1)/2 \rfloor$, i.e. $\mathbf{r} = \mathbf{c} + \mathbf{e}$ for a unique $\mathbf{c} \in C$ and the *error vector* $\mathbf{e}$ with $\text{wt}(\mathbf{e}) \leq e$. Such decoding is called *bounded (up to half the minimum distance) decoding* . If we want to go beyond the error-correcting capacity, we have to give up the uniqueness. In this case *nearest codeword* decoder returns the list of codewords nearest to the given received word. More discussion on this topic is in Section 1.3.4.

There are two basic methods for decoding linear codes. The first one is *exhaustive search* , namely one measures the distance from the received word $\mathbf{r}$ to every codeword from C. Since a q-ary $[n,k]$ code C has q^k codewords, the complexity of such method is at most nq^k and is $nq^k/2$ on average. The second method uses the parity check matrix H of the code C. Recall that the code C is characterized as a null-space of H. Now let $\mathbf{r} = \mathbf{c} + \mathbf{e}$. Then we can compute $\mathbf{s}(\mathbf{r}) = H\mathbf{r}^T = H(\mathbf{c}^T + \mathbf{e}^T) = H\mathbf{e}^T$. $\mathbf{s}(\mathbf{r})$ is called the *syndrome* of $\mathbf{r}$. As we see that the syndrome of $\mathbf{r}$ coincides with the one of an error vector $\mathbf{e}$. This means that $\mathbf{r}$ and $\mathbf{e}$ lie in one coset of $\mathbb{F}_q^n/C$. So according to maximum-likelihood decoding one just has to choose vectors of the smallest weight in each coset (they are called *coset leaders*), then when $\mathbf{r}$ is received, one checks using $\mathbf{s}(\mathbf{r})$ in which coset does $\mathbf{r}$ lie, picks the coset leader $\mathbf{e}$ of that coset, and decodes as $\mathbf{r} - \mathbf{e}$. So one has to store q^{n-k} coset leaders, time complexity then is negligible. This method is useful when the redundancy $n-k$ is low. The above method is referred to as *syndrome decoding*. More on syndrome decoding in Section 1.3.4.

Another idea proves to be very effective, we sketch it now (cf. [13]). Let $\mathbf{r} = \mathbf{c} + \mathbf{e}$ with $\mathbf{c} \in C$ and $\text{wt}(\mathbf{e}) = t \leq e$. Take randomly some subset I of $\{1, \ldots, n\}$ of cardinality k. Assume that the errors occurred outside I, i.e. $e_i = 0, i \in I$ and $r_i = c_i, i \in I$. Since k positions of $\mathbf{r}$ now assumed to be known, we are left with determining the remaining $n - k$. These can be recovered from the equations $H\mathbf{r}^T = \mathbf{s}(\mathbf{r})$. Having k positions of $\mathbf{r}$ we are left with $(n-k) \times (n-k)$ linear system. The corresponding submatrix of H has full-rank with high probability if the code C is random. Solving this linear system one obtains the values $e_i, i \notin I$. One can then check whether $H(\mathbf{r} - \mathbf{e})^T = 0$ with so obtained $\mathbf{e}$. If $H(\mathbf{r} - \mathbf{e})^T \neq 0$, the next set I is chosen. The probability of the fact that indeed the errors do not occur at the positions marked by I is $\binom{n-k}{t}/\binom{n}{t}$. Therefore, on average one needs to perform $\binom{n}{t}/\binom{n-k}{t}$ trials before guessing correctly. The procedure we have just sketched is called *covering set decoding* using the parity check matrix H. There exists also a variant based on the generator matrix G and the so-called *information sets* . The complexity at every trial step is negligible, so the expected number of trials determines the total complexity. Asymptotically, covering sets decoding is better than either exhaustive search or syndrome decoding.

1.1. CODES AND DECODING PROBLEM

Code-base cryptography

At the moment public key encryption schemes based on number theoretic problems like factorization (RSA) or discrete logarithm problem, e.g. in a group of points of an elliptic curve (ECC) are widely used. This may change with an appearance of a quantum computer, since then the so-called Shor's algorithm [117] that works on a quantum computer is able to solve both the factorization and the discrete logarithm problem efficiently. Although the time and even the fact of possibility of a construction of a quantum computer are quite debatable, it makes a lot of sense to start to look for alternatives now. One of the alternatives is the code-base cryptography (cf. [104]). By this one understands constructing different cryptographic primitives using (linear) codes. We refer to [104] for an overview of the subject.

Now let us briefly describe a public key cryptosystem due to McEliece [97]. The system can be described as follows.

Secret key: A $k \times n$ generator matrix G of a code over $\mathbb{F}_q$, for which an efficient algorithm correcting t errors is known [1]. An $k \times k$ non-singular "scrambler" matrix S and an $n \times n$ permutation matrix P.

Public key: The $k \times n$ matrix $G' = SGP$. The field $\mathbb{F}_q$ (together with the description), parameters n, k, t.

Encryption: To encrypt a message $\mathbf{m} \in \mathbb{F}_q^k$ with the public key G' one chooses a random vector $\mathbf{e} \in \mathbb{F}_q^n$ with $\mathrm{wt}(\mathbf{e}) \leq t$ and sends $\mathbf{r} = \mathbf{m}G' + \mathbf{e}$.

Decryption: Upon receiving $\mathbf{r} = \mathbf{m}SGP + \mathbf{e} \in \mathbb{F}_q^n$ one can retrieve the message $\mathbf{m}$ as follows. First using the secret matrix P compute $\mathbf{r}P^{-1} = \mathbf{m}SG + \mathbf{e}P^{-1}$, where $\mathrm{wt}(\mathbf{e}P^{-1}) \leq t$ as before. Now using the decoding algorithm for the code generated by G one can correct t errors and obtain the intermediate message $\mathbf{m}S$. By multiplying by S on the right one finally retrieves the initial plaintext message $\mathbf{m}$.

The idea of the McEliece construction is to make the "good" code generated by G look like a "random" code generated by G'. It is then supposed that an attacker has nothing to do better than just trying to decode t errors using a "random" code. In general this problem is hard (NP-hard). One may try to undertake some generic attacks based on the ideas from the previous subsection. Since in this chapter we address the decoding method for general linear codes, it may be relevant also to consider our work in this context.

---

[1] McEliece used the Goppa codes. Goppa-based cryptosystem is considered to be secure so far.

1.2 Gröbner bases for system solving

The theory of Gröbner basis is about solving systems of polynomial equations in several variables and can be viewed as a common generalization of Gaussian elimination in linear algebra that deals with linear systems of equations in several variables and the Euclidean algorithm that is about polynomial equations of arbitrary degree in one variable. In this subsection we give a brief overview of monomial orders, Gröbner bases and their use in polynomial system solving. This subsection is only intended to refresh these notions; for a thorough exposition of the material the reader can use e.g [55, 90, 75].

Let $\mathbb{F}$ be a field and let $\mathbb{F}[X_1, \ldots, X_n] = \mathbb{F}[X]$ be the polynomial ring in n variables over $\mathbb{F}$. In commutative algebra objects like polynomials, ideals, quotients are intensively studied. If we want to do computations with these objects we must somehow impose an order on them, so that we know which way a computation will go. Let $Mon(X)$ be the set of all monomials in the variables $X = (X_1, \ldots, X_n)$.

Definition 1.2.1 A *monomial order* on $\mathbb{F}[X]$ is any relation $>$ on $Mon(X)$ such that

1. $>$ is a total order on $Mon(X)$,

2. $>$ is multiplicative, i.e. $X^\alpha > X^\beta$ implies $X^\alpha \cdot X^\gamma > X^\beta \cdot X^\gamma$ for all vectors γ with non-negative integer entries; here $X^\alpha = X_1^{\alpha_1} \cdot \ldots \cdot X_n^{\alpha_n}$,

3. $>$ is a well-order, i.e. every non-empty subset of $Mon(X)$ has a minimal element.

Example 1.2.2 Here are some orders that will be used in this thesis.

- *Lexicographic order* induced by $X_1 > \cdots > X_n : X^\alpha >_{lp} X^\beta$ if and only if there exists an s such that $\alpha_1 = \beta_1, \ldots, \alpha_{s-1} = \beta_{s-1}, \alpha_s > \beta_s$.

- *Degree reverse lexicographic order* induced by $X_1 > \cdots > X_n : X^\alpha >_{dp} X^\beta$ if and only if $|\alpha| := \alpha_1 + \cdots + \alpha_n > \beta_1 + \cdots + \beta_n =: |\beta|$ or if $|\alpha| = |\beta|$ and there exists an s such that $\alpha_n = \beta_n, \ldots, \alpha_{n-s+1} = \beta_{n-s+1}, \alpha_{n-s} < \beta_{n-s}$.

- *Block order* or *product order* . Let X and Y be two ordered sets of variables, $>_1$ a monomial order on $\mathbb{F}[X]$ and $>_2$ a monomial order on $\mathbb{F}[Y]$. The block order on $\mathbb{F}[X, Y]$ is the following: $X^{\alpha_1} Y^{\beta_1} > X^{\alpha_2} Y^{\beta_2}$ if and only if $X^{\alpha_1} >_1 X^{\alpha_2}$ or if $X^{\alpha_1} =_1 X^{\alpha_2}$ and $Y^{\beta_1} >_2 Y^{\beta_2}$.

1.2. GRÖBNER BASES FOR SYSTEM SOLVING

Definition 1.2.3 Let $>$ be a monomial order on $\mathbb{F}[X]$. Let $f = \sum_\alpha c_\alpha X^\alpha$ be a non-zero polynomial from $\mathbb{F}[X]$. Let α_0 be such that $c_{\alpha_0} \neq 0$ and $X^{\alpha_0} > X^\alpha$ for all $\alpha \neq \alpha_0$ with $c_\alpha \neq 0$. Then $\mathrm{lc}(f) := c_{\alpha_0}$ is called the *leading coefficient* of f, $\mathrm{lm}(f) := X^{\alpha_0}$ is called the *leading monomial* of f, $\mathrm{lt}(f) := c_{\alpha_0} X^{\alpha_0}$ is called the *leading term* of f, moreover $\mathrm{tail}(f) := f - \mathrm{lt}(f)$.

Having these notions we are ready to define the notion of a Gröbner basis.

Definition 1.2.4 Let I be an ideal in $\mathbb{F}[X]$. The *leading ideal* of I with respect to $>$ is defined as $L_>(I) := \langle \mathrm{lt}(f) | f \in I, f \neq 0 \rangle$. The $L_>(I)$ is sometimes abbreviated by $L(I)$. A monomial m is called a *standard monomial* for an ideal I, if $m \notin L(I)$. A finite subset $G = \{g_1, \ldots, g_m\}$ of I is called a *Gröbner basis* for I with respect to $>$ if $L_>(I) = \langle \mathrm{lt}(g_1), \ldots, \mathrm{lt}(g_m) \rangle$.

Example 1.2.5 Consider two polynomials $f = X^3, g = Y^4 - X^2Y$ from $\mathbb{F}[X,Y]$, where $\mathbb{F}$ is any field. We claim that f and g constitute a Gröbner basis of an ideal $I = \langle f, g \rangle$ with respect to the degree reverse lexicographic order $>_{dp}$ with $X > Y$. For this we need to show that $L(I) = \langle \mathrm{lt}(f), \mathrm{lt}(g) \rangle$. We have $\mathrm{lt}(f) = X^3$ and $\mathrm{lt}(g) = Y^4$. Thus we have to show that $\mathrm{lt}(h)$ is divisible either by X^3 or by Y^4, for any $h \in I$. A polynomial h can be written as $h = af + bg = aX^3 + b(Y^4 - X^2Y)$. If $\deg(a) > 1 + \deg(b)$, then $\mathrm{lm}(h) = \mathrm{lm}(a)X^3$. If $\deg(a) < 1 + \deg(b)$, then $\mathrm{lm}(h)$ is divisible by Y^4. If $\deg(a) = 1 + \deg(b)$ and $\mathrm{lm}(a)X^3 \neq \mathrm{lm}(b)Y^4$, then $\mathrm{lm}(h) = \mathrm{lm}(a)X^3$. If $\deg(a) = 1 + \deg(b)$ and $\mathrm{lm}(a)X^3 = \mathrm{lm}(b)Y^4$, then $\mathrm{lm}(h)$ is divisible by X^3.

Every ideal has a Gröbner basis. By doing some additional operations on the elements of a Gröbner basis, one can construct a *reduced* Gröbner basis. It has the property that if $G = \{g_1, \ldots, g_m\}$ is a reduced Gröbner basis, then no $\mathrm{lm}(g_i)$ divides any monomial of $\mathrm{tail}(g_j), j \neq i$. The reduced Gröbner basis of an ideal with respect to a given monomial order is unique.

There are several algorithms for computing Gröbner bases. Historically the first is Buchberger's algorithm [26] and its numerous improvements and optimizations are implemented in several computer algebra systems like for example SINGULAR [74], MAGMA [38], and CoCoA [49]. Also there are algorithms F4 and F5 [63, 64]. The algorithm F4 is implemented e.g. in MAGMA and FGB [66].

For solving systems of polynomial equations with the use of Gröbner bases we need the so-called elimination orders.

Definition 1.2.6 Let S be some subset of variables in X. A monomial order $>$ on $\mathbb{F}[X]$ is called an *elimination order* with respect to S if for all $f \in \mathbb{F}[X]$ from the fact that $\mathrm{lm}(f) \in \mathbb{F}[X \setminus S]$ follows that $f \in \mathbb{F}[X \setminus S]$.

For example, let $>$ be the block order $(>_1, >_2)$ on $\mathbb{F}[S,T]$ ($S \subset X$ and $T = X \setminus S$), where $>_1$ is defined on $\mathbb{F}[S]$ and $>_2$ is defined on $\mathbb{F}[T]$. Then $>$ is an elimination order with respect to S. In particular, lexicographic order is an elimination order with respect to any subset S of X. Due to this property of the lexicographic order we have the following theorem that can be obtained from the Elimination Theorem, p.114 and the theorem about finiteness, p.232, [55]; also p.83 [90].

Theorem 1.2.7 *Let $f_1(X) = \cdots = f_m(X) = 0$ be a system of polynomial equations defined over $\mathbb{F}[X]$ with $X = (X_1, \ldots, X_n)$, such that it has finitely many solutions in $\overline{\mathbb{F}}^n$, where $\overline{\mathbb{F}}$ is the algebraic closure of $\mathbb{F}$. Let $I = \langle f_1, \ldots, f_m \rangle$ be an ideal defined by the polynomials in the system and let G be a Gröbner basis for I with respect to $>_{lp}$ induced by $X_n < \cdots < X_1$. Then there are elements $g_1, \ldots, g_n \in G$ such that*

$$\begin{cases} g_n \in \mathbb{F}[X_n], & lt(g_n) = c_n X_n^{m_n}, \\ g_{n-1} \in \mathbb{F}[X_{n-1}, X_n], & lt(g_{n-1}) = c_{n-1} X_{n-1}^{m_{n-1}}, \\ \cdots \\ g_1 \in \mathbb{F}[X_1, \ldots, X_n], & lt(g_1) = c_1 X_1^{m_1}. \end{cases}$$

for some positive integers $m_i, i = 1, \ldots, n$ and elements $c_i \in \mathbb{F}^*, i = 1, \ldots, n$.

It is clear how to solve the system I now. After computing G, first solve a univariate equation $g_n(X_n) = 0$. Let $a_1^{(n)}, \ldots, a_{l_n}^{(n)}$ be the roots. For every $a_i^{(n)}$ then solve $g_{n-1}(X_{n-1}, a_i^{(n)}) = 0$ to find possible values for X_{n-1}. Repeat this process until all the coordinates of all candidate solutions are found. The candidates form a finite set $Can \subseteq \overline{\mathbb{F}}^n$. Test all other elements of G on whether they vanish at elements of Can. If there is some $g \in G$ that does not vanish at some $can \in Can$, then discard can from Can. Since the number of solutions is finite the above procedure terminates. For a concrete example see Example 1.3.4. Systems with finitely many solutions in the algebraic closure are referred to as zero-dimensional systems.

Remark 1.2.8 Usually from the practical point of view finding a Gröbner basis with respect to an elimination order is harder than with respect to some degree-refining order, like the degree reverse lexicographic order. Therefore, a conversion technique like FGLM [65] comes in hand here. It enables one to convert a basis with respect to one order, for instance some degree-refining order, to another one, such as the lexicographic order. For solving, we actually need an elimination order, but sometimes it is possible to obtain a result with a degree (in fact any) order, see e.g. Theorem 1.4.45.

1.2. GRÖBNER BASES FOR SYSTEM SOLVING

Some technical details on computation of Gröbner bases follow. The *s-polynomial* of $f, g \in \mathbb{F}[X] \setminus \{0\}$ with leading monomials $\mathrm{lm}(f) = X^\alpha$ and $\mathrm{lm}(g) = X^\beta$ is defined by

$$\mathrm{spoly}(f, g) = X^{\gamma-\alpha} f - \frac{\mathrm{lc}(f)}{\mathrm{lc}(g)} X^{\gamma-\beta} g,$$

where $\gamma = (\max(\alpha_1, \beta_1), \ldots, \max(\alpha_n, \beta_n))$.

Definition 1.2.9 [Standard representation, reduced normal form] Let $f, g_1, \ldots, g_m \in P$, and let $h_1, \ldots, h_m \in P$. Then

$$f = \sum_{i=1}^{m} h_i \cdot g_i,$$

is called a *standard representation* of f w.r.t $g_1, \ldots, g_m$, if for all i either $h_i \cdot g_i = 0$, or $\mathrm{lm}(h_i \cdot g_i) \leq \mathrm{lm}(f)$.

Definition 1.2.10 A polynomial f is said to be reduced against a generating system G, if and only if $\mathrm{supp}(f) \cap \mathrm{L}(G) = \emptyset$. A polynomial r is called the *reduced normal form* of f against G, if and only if it is reduced against G and $f - r$ has standard representation with respect to G. As the reduced normal form is unique, we may denote this by $r = \mathrm{NF}(f, G)$.

Theorem 1.2.11 (Buchberger's criterion, cf. Theorem 1.7.3 [75]) Let I be an ideal from $\mathbb{F}[X_1, \ldots, X_n]$ and $G = \{g_1, \ldots, g_m\} \subset I$. The following are equivalent:

1. G is a Gröbner basis of I.
2. $\mathrm{NF}(f|G) = 0$ for all $f \in I$.
3. Each $f \in I$ has a standard representation with respect to G.
4. G generates I and $\mathrm{NF}(\mathrm{spoly}(g_i, g_j)|G) = 0$ for $i, j = 1, \ldots, m$.
5. G generates I and $\mathrm{NF}(\mathrm{spoly}(g_i, g_j)|G_{ij}) = 0$ for a suitable subset $G_{ij} \subset G$ and $i, j = 1, \ldots, m$.

Proposition 1.2.12 (Product criterion, cf. e.g. p.63, [75]) Let $f, g \in \mathbb{F}[X_1, \ldots, X_n]$ be polynomials such that $\mathrm{lcm}(\mathrm{lm}(f), \mathrm{lm}(g)) = \mathrm{lm}(f) \cdot \mathrm{lm}(g)$, then

$$\mathrm{NF}(\mathrm{spoly}(f, g)|\{f, g\}) = 0.$$

In particular this holds if $\mathrm{lm}(f)$ and $\mathrm{lm}(g)$ are coprime.

In Chapter 2 we work primarily over the field $\mathbb{F}_2$. We will be interested in doing computations in the ring $\mathbb{F}_2[X_1, \ldots, X_n]/\langle X_1^2 + X_1, \ldots, X_n^2 + X_n\rangle$. The POLYBORIframework is designed for Gröbner basis computations with the so-called *Boolean polynomials* as canonical representatives of residue classes in $\mathbb{F}_2[X_1, \ldots, X_n]/\langle X_1^2 + X_1, \ldots, X_n^2 + X_n\rangle$. Therefore, we will consider Boolean polynomials as the polynomials from $\mathbb{F}_2[X_1, \ldots, X_n]$ such that every variable x_i occurs with exponent 0 or 1. In [25] it is described, how to apply classical Gröbner basis theory for polynomial rings to this quotient ring. On the description and principles of the POLYBORIframework the reader is referred to [24]. The following criterion will play a crucial role in explaining our construction in Section 2.4.1.

Proposition 1.2.13 (Linear lead criterion, cf. [24]) *Let $f \in \mathbb{F}_2[X_1, \ldots, X_n]$ be a Boolean polynomial such that $f = l \cdot g$, $lm(l) = X_i$ for some i and g any Boolean polynomial, then* $\mathrm{NF}(\mathrm{spoly}(f, X_i^2 + X_i)|\{f, X_1^2 + X_1, \ldots, X_n^2 + X_n\}) = 0$.

The case $g = 1$ will be of interest for us.

1.3 Gröbner bases in coding theory: overview

This chapter is devoted to decoding and finding the minimum distance of arbitrary linear codes with the use of Gröbner bases. In recent years a lot of attention was paid to this question for cyclic codes, which form a particular subclass of linear codes. In this section we give a survey on some existing methods for decoding and finding the minimum distance with Gröbner bases.

Quite a lot of methods exist for decoding cyclic codes and the literature on this topic is vast. We just mention [3, 15, 96, 108, 126]. But all these methods do not correct up to the true error-correcting capacity. The theory of Gröbner bases is used to remedy this problem. These methods are roughly divided into the following categories:

- Newton identities method [5, 6, 7, 8, 9, 17, 18, 40]

- Power sums method or Cooper's philosophy [37, 39, 40, 41, 51, 113].

The term "Cooper's philosophy" was initially used during the talk [99]. In Sections 1.1 and 1.2 necessary background on linear codes and Gröbner bases is given, as well as the notation we are going to use throughout the thesis. In Subsections 1.3.1 and 1.3.2 we give an overview of the methods based on power sums and Newton identities together with examples. Subsection 1.3.3 is devoted to the case of arbitrary linear codes. Namely, we look at the

1.3. GRÖBNER BASES IN CODING THEORY: OVERVIEW

method of Fitzgerald and Lax. We should mention that there exist other Gröbner bases-based methods for arbitrary linear codes, e.g. generalizations of the Cooper's philosophy [72, 115], applications of Padé approximation [61, 62, 68], FGLM-like techniques [19, 20, 91], key equation [67]. These methods are out of scope of this thesis. We made an extensive bibliography, so that the reader is able to look at numerous sources that exist in the area.

The idea in polynomial system solving methods for decoding codes is as follows. One associates to the code (and a received vector) certain polynomial system over a finite field. The solution(s) of such a system should yield the corresponding error vector(s) in more or less straightforward way. One distinguishes between two concepts:

- *Generic decoding* : Solve some system $S(C)$ and obtain some "closed formulas" F. Evaluating these formulas at data specific to a received word **r** should yield a solution to the decoding problem. For example for $f \in F : f(syndrome(\mathbf{r}), x) = poly(x)$. The roots of $poly(x) = 0$ yield error positions. This is something that is called *general errorlocator polynomial f*. The case of the cyclic codes is covered in the following material.

- *Online decoding* : Solve some system $S(C, \mathbf{r})$. The solutions should solve the decoding problem.

One can ask about computational effort in the two scenarios above.

- Generic decoding. Preprocessing: very hard, since one has to operate with really huge symbolic data to obtain "closed formulas". Decoding: relatively simple, but only if the closed formulas are sparse. Otherwise, evaluating the closed formulas in received-word-specific data is too cumbersome.

- Online decoding. Preprocessing: none. Decoding: hard, since one has to solve a non-linear system of equations.

1.3.1 Cooper's philosophy and its development

Throughout this and the next subsection we denote the field with q elements as $\mathbb{F}_q$ and the splitting field of $X^n - 1$ over $\mathbb{F}_q$ as $\mathbb{F}$. Denote by m the degree of $\mathbb{F}$ over $\mathbb{F}_q$ and by a some primitive n-th root of unity. We assume further that $(n, q) = 1$. In this subsection we give an overview of the so-called Cooper's philosophy or the power sums method. The idea here is basically to write parity check equations with unknowns for error positions and error values and then try to solve with respect to these unknowns by adding some natural restrictions on them.

If i is in the defining set of C, then
$$(1, a^i, \ldots, a^{(n-1)i})\mathbf{c}^T = c_0 + c_1 a^i + \cdots + c_{n-1} a^{(n-1)i} = c(a^i) = 0.$$
Hence $(1, a^i, \ldots, a^{(n-1)i})$ is a parity check of C. Let $\{i_1, \ldots, i_{n-k}\}$ be a defining set of C. Then a parity check matrix H of C can be represented as a matrix with entries in $\mathbb{F}$:

$$H = \begin{pmatrix} 1 & a^{i_1} & a^{2i_1} & \cdots & a^{(n-1)i_1} \\ 1 & a^{i_2} & a^{2i_2} & \cdots & a^{(n-1)i_2} \\ \vdots & \vdots & \vdots & \ddots & \vdots \\ 1 & a^{i_{n-k}} & a^{2i_{n-k}} & \cdots & a^{(n-1)i_{n-k}} \end{pmatrix}. \tag{1.1}$$

Let $\mathbf{c}$, $\mathbf{r}$ and $\mathbf{e}$ be the transmitted codeword, the received word and the error vector, respectively. Then $\mathbf{r} = \mathbf{c} + \mathbf{e}$. Denote the corresponding polynomials by $c(x)$, $r(x)$ and $e(x)$, respectively. If we apply the parity check matrix to $\mathbf{r}$, we obtain
$$\mathbf{s}^T := H\mathbf{r}^T = H(\mathbf{c}^T + \mathbf{e}^T) = H\mathbf{c}^T + H\mathbf{e}^T = H\mathbf{e}^T,$$
since $H\mathbf{c}^T = 0$, where $\mathbf{s}$ is the so-called *syndrome vector*. Define $s_i = r(a^i)$ for all $i = 1, \ldots, n$. Then $s_i = e(a^i)$ for all i in the complete defining set, and these s_i are called the *known syndromes*. The remaining s_i are called the *unknown syndromes*. We have that the vector $\mathbf{s}$ above has entries $\mathbf{s} = (s_{i_1}, \ldots, s_{i_{n-k}})$. Let t be the number of errors that occurred while transmitting $\mathbf{c}$ over a noisy channel. If the error vector is of weight t, then it is of the form
$$\mathbf{e} = (0, \ldots, 0, e_{j_1}, 0, \ldots, 0, e_{j_l}, 0, \ldots, 0, e_{j_t}, 0, \ldots, 0),$$
more precisely there are t indices j_l with $1 \leq j_1 < \cdots < j_t \leq n$ such that $e_{j_l} \neq 0$ for all $l = 1, \ldots, t$ and $e_j = 0$ for all j not in $\{j_1, \ldots, j_t\}$. We obtain
$$s_{i_u} = r(a^{i_u}) = e(a^{i_u}) = \sum_{l=1}^{t} e_{j_l}(a^{i_u})^{j_l}, \quad 1 \leq u \leq n - k. \tag{1.2}$$

The $a^{j_1}, \ldots, a^{j_t}$ but also the $j_1, \ldots, j_t$ are called the *error locations*, and the $e_{j_1}, \ldots, e_{j_t}$ are called the *error values*. Define $z_l = a^{j_l}$ and $y_l = e_{j_l}$. Then $z_1, \ldots, z_t$ are the error locations and $y_1, \ldots, y_t$ are the error values and the syndromes in (1.2) become *generalized power sum functions*

$$s_{i_u} = \sum_{l=1}^{t} y_l z_l^{i_u}, \quad 1 \leq u \leq n - k. \tag{1.3}$$

1.3. GRÖBNER BASES IN CODING THEORY: OVERVIEW

In the binary case the error values are $y_i = 1$, and the syndromes are the ordinary power sums.

Now we give a description of *Cooper's philosophy*. As the receiver does not know how many errors occurred, the upper bound t is replaced by the error-correcting capacity e and some z_l's are allowed to be zero, while assuming that the number of errors is at most e. The following variables are introduced: $X_1, \ldots, X_{n-k}, Z_1, \ldots, Z_e$ and $Y_1, \ldots, Y_e$, where X_u stands for the syndrome $s_{i_u}, 1 \leq u \leq n-k$; Z_l stands for the error location z_l for $1 \leq l \leq t$, and 0 for $t < l \leq e$; and finally Y_l stands for the error value y_l for $1 \leq l \leq t$, and any element of $\mathbb{F}_q \setminus \{0\}$ for $t < l \leq e$. The syndrome equations (1.2) are rewritten in terms of these variables as power sums:

$$f_u := \sum_{l=1}^{e} Y_l Z_l^{i_u} - X_u = 0, \quad 1 \leq u \leq n-k.$$

We also add some other equations in order to specify the range of values that can be achieved by our variables, namely:

$$\epsilon_u := X_u^{q^m} - X_u = 0, \quad 1 \leq u \leq n-k,$$

since $s_j \in \mathbb{F}$;

$$\eta_l := Z_l^{n+1} - Z_l = 0, \quad 1 \leq l \leq e,$$

since a^{j_l} are either n-th roots of unity or zero; and

$$\lambda_l := Y_l^{q-1} - 1 = 0, \quad 1 \leq l \leq e,$$

since $y_l \in \mathbb{F}_q \setminus \{0\}$. We obtain the following set of polynomials in the variables $X = (X_1, \ldots, X_{n-k})$, $Z = (Z_1, \ldots, Z_e)$ and $Y = (Y_1, \ldots, Y_e)$:

$$F_C = \{f_u, \epsilon_u, \eta_l, \lambda_l : 1 \leq u \leq n-k, 1 \leq l \leq e\} \subset \mathbb{F}_q[X, Z, Y]. \quad (1.4)$$

The zero-dimensional ideal I_C generated by F_C is called the *CRHT-syndrome ideal* associated to the code C, and the variety $V(F_C)$ defined by F_C is called the *CRHT-syndrome variety*, after Chen, Reed, Helleseth and Truong, see [39, 40, 41]. We have $V(F_C) = V(I_C)$.

Initially decoding of cyclic codes was essentially brought to finding the reduced Gröbner basis of the CRHT-ideal. One then had to look through the entire reduced Gröbner bases, which usually is quite complicated, in order to extract elements needed for decoding. It turns out that adding more polynomials to this ideal gives better results [113]. By adding polynomials

$$\chi_{l,m} := Z_l Z_m p(n, Z_l, Z_m) = 0, \quad 1 \leq l < m \leq e$$

to F_C, where

$$p(n, X, Y) = \frac{X^n - Y^n}{X - Y} = \sum_{i=0}^{n-1} X^i Y^{n-1-i}, \qquad (1.5)$$

we ensure that for all l and m either Z_l and Z_m are distinct or at least one of them is zero. The resulting set of polynomials in $\mathbb{F}_q[X, Z, Y]$:

$$F'_C := \{f_u, \epsilon_u, \eta_i, \lambda_i, \chi_{l,m} : 1 \leq u \leq n - k, 1 \leq i \leq e, 1 \leq l < m \leq e\}. \quad (1.6)$$

The ideal generated by F'_C is denoted by I'_C. By investigating the structure of I'_C and its reduced Gröbner basis with respect to lexicographic order induced by $X_1 < \cdots < X_{n-k} < Z_e < \cdots < Z_1 < Y_1 < \cdots < Y_e$, the following result is proved, see [113][Theorem 6.8, 6.9].

Theorem 1.3.1 *Every cyclic code C possesses a general error-locator polynomial L_C. This means that there exists a unique polynomial L_C from $\mathbb{F}_q[X_1, \ldots, X_{n-k}, Z]$ that satisfies the following two properties:*

- *$L_C = Z^e + a_{t-1} Z^{e-1} + \cdots + a_0$ with $a_j \in \mathbb{F}_q[X_1, \ldots, X_{n-k}]$, $0 \leq j \leq e-1$;*

- *given a syndrome $\mathbf{s} = (s_{i_1}, \ldots, s_{i_{n-k}}) \in \mathbb{F}^{n-k}$ corresponding to an error of weight $t \leq e$ and error locations $\{k_1, \ldots, k_t\}$, if we evaluate the $X_u = s_{i_u}$ for all $1 \leq u \leq n - k$, then the roots of $L_C(\mathbf{s}, Z)$ are exactly $a^{k_1}, \ldots, a^{k_t}$ and 0 of multiplicity $e - t$, in other words*

$$L_C(\mathbf{s}, Z) = Z^{e-t} \prod_{i=1}^{t} (Z - a^{k_i}).$$

Such an error-locator polynomial actually is an element of the reduced Gröbner basis of I'_C. Having this polynomial, decoding of the cyclic code C reduces to univariate factorization. The main effort here is finding the reduced Gröbner basis of I'_C. In general this is infeasible already for moderate size codes, for small codes, though, it is possible to apply this technique successfully [114].

Example 1.3.2 As an example we consider finding the general error-locator polynomial for a binary cyclic BCH code C with parameters [15,7,5] that corrects 2 errors. This code has $\{1, 3\}$ as a defining set. So here $q = 2, m = 4$, and $n = 15$. The field $\mathbb{F}_{16}$ is the splitting field of $X^{15} - 1$ over $\mathbb{F}_2$. During this example we show how the idea of the Cooper's philosophy is applied. For rigorous justification of the steps below, see [39, 40, 41, 113, 114]. Note that we may write the equations only for the elements from the defining set $\{1, 3\}$

1.3. GRÖBNER BASES IN CODING THEORY: OVERVIEW

as all the others are just consequences of those. Following the description above we write generators F'_C of the ideal I'_C in the ring $\mathbb{F}_{16}[X_1, X_2, Z_1, Z_2]$:

$$\begin{cases} Z_1 + Z_2 - X_1, & Z_1^3 + Z_2^3 - X_2, \\ X_1^{16} - X_1, & X_2^{16} - X_2, \\ Z_1^{16} - Z_1, & Z_2^{16} - Z_2, \\ Z_1 Z_2 p(15, Z_1, Z_2). \end{cases}$$

We suppress the equations λ_1 and λ_2 as error values are over $\mathbb{F}_2$. In order to find the general error locator polynomial we compute the reduced Gröbner basis G of the ideal I'_C with respect to the lexicographical order induced by $X_1 < X_2 < Z_2 < Z_1$. The elements of G are:

$$\begin{cases} X_1^{16} + X_1, \\ X_2 X_1^{15} + X_2, \\ X_2^8 + X_2^4 X_1^{12} + X_2^2 X_1^3 + X_2 X_1^6, \\ Z_2 X_1^{15} + Z_2, \\ Z_2^2 + Z_2 X_1 + X_2 X_1^{14} + X_1^2, \\ Z_1 + Z_2 + X_1 \end{cases}$$

According to Theorem 6.8 (cf. [113]) the general error correcting polynomial L_C is then a unique element of G of degree 2 with respect to Z_2. So $L_C \in \mathbb{F}_2[X_1, X_2, Z]$ is

$$L_C(X_1, X_2, Z) = Z^2 + ZX_1 + X_2 X_1^{14} + X_1^2.$$

Let us see how decoding using L_C works. Let $\mathbf{r} = (0, 1, 1, 0, 0, 1, 0, 1, 0, 1, 0, 0, 1, 1, 0, 1)$ be a received word with at most 2 errors. In the field $\mathbb{F}_{16}$ with a primitive element a, such that $a^4 + a + 1 = 0$, a is also a 15-th root of unity. Then the syndromes are $\mathbf{s}_1 = a^2, \mathbf{s}_3 = a$. Plug them into L_C in place of X_1 and X_2 and obtain:

$$L_C(Z) = Z^2 + a^2 Z + a(a^2)^{14} + (a^2)^2 = Z^2 + a^2 Z + a^9.$$

Factorizing yields $L_C = (Z + a^3)(Z + a^6)$. According to Theorem 1.3.1, exponents 3 and 6 show exactly the error locations minus 1. So the errors occurred on positions 4 and 7.

Consider another example. Let $\mathbf{r} = (0, 0, 0, 1, 0, 0, 0, 1, 1, 1, 1, 1, 0, 0, 0)$ be the received word with at most 2 errors. The syndromes are now $\mathbf{s}_1 = a^8, \mathbf{s}_3 = a^9$. Plug them into L_C in place of X_1 and X_2 and obtain:

$$L_C(Z) = Z^2 + a^8 Z + a^9(a^8)^{14} + (a^8)^2 = Z^2 + a^8 Z.$$

Factorizing yields $L_C = Z(Z + a^8)$. Thus 1 error occurred according to Theorem 1.3.1, namely on position 8+1=9.

This method can be adapted to correct erasures [113], and to find the minimum distance of a code [116]. The basic approach is as follows. We are working again with the cyclic code C with parameters $[n, k, d]$ over $\mathbb{F}_q$. Let $w \leq d$. Denote by $J_C(w)$ the set of equations (1.6) for $t = w$ and variables X_i assigned to zero and the equations $Z_i^{n+1} - Z_i = 0$ replaced by $Z_i^n - 1 = 0$. In the binary case we have the following result that can be deduced from Theorem 3.3 and Corollary 3.4 [116]:

Theorem 1.3.3 *Let C be a binary $[n, k, d]$ cyclic code with $S_C = \{i_1, \ldots, i_v\}$ as defining set. Let $1 \leq w \leq n$ and let $J_C(w)$ denote the system:*

$$\begin{cases} Z_1^{i_1} + \cdots + Z_w^{i_1} = 0, \\ \vdots \\ Z_1^{i_v} + \cdots + Z_w^{i_v} = 0, \\ Z_1^n - 1 = 0, \\ \vdots \\ Z_w^n - 1 = 0, \\ p(n, Z_i, Z_j) = 0, 1 \leq i < j \leq w. \end{cases}$$

Then the number of solutions of $J_C(w)$ is equal to $w!$ times the number of codewords of weight w. And for $1 \leq w \leq d$:

- *either $J_C(w)$ has no solution, which is equivalent to $w < d$,*

- *or $J_C(w)$ has a solution, which is equivalent to $w = d$.*

So, the method of finding the minimum distance is based on replacing syndrome variables by zeros and then searching for solutions of corresponding parametrized systems. In the previous theorem $J_C(w)$ is parametrized by w. We also mention the notion of *accelerator polynomials*. The idea is as follows. Since, when trying to find the minimum distance of a code we are only interested in the question whether the corresponding system $J_C(w)$ has solutions or not, we may add some polynomials $A_C(w)$ with the property that if we enlarge our system $J_C(w)$ with these polynomials and the system $J_C(w)$ had some solutions, then the new system $A_C(w) \cup J_C(w)$ also has some solutions. So not all solutions are lost. In [100] it is shown, how to choose such polynomials $A_C(w)$, so that solving the system $A_C(w) \cup J_C(w)$ takes less time, than solving $J_C(w)$.

It is possible to adapt the method to finding codewords of certain weights, and thus the weight enumerator of a given code.

1.3. GRÖBNER BASES IN CODING THEORY: OVERVIEW

Example 1.3.4 As an example application of Theorem 1.3.3 we show how to determine the minimum distance of a cyclic code C from Example 1.3.2. This binary cyclic code C has parameters [15,7] and has a defining set $\{1,3\}$, so the assumptions of Theorem 1.3.3 are satisfied. We have to look at all systems $J_C(w)$ starting from $w = 1$, until we encounter a system, which has some solutions. The system $J_C(w)$ is

$$\begin{cases} Z_1 + \cdots + Z_w = 0, \\ Z_1^3 + \cdots + Z_w^3 = 0, \\ Z_1^{15} - 1 = 0, \\ \vdots \\ Z_w^{15} - 1 = 0, \\ p(15, Z_i, Z_j) = 0, 1 \leq i < j \leq w. \end{cases}$$

For $w = 1, \ldots, 4$ the reduced Göbner basis of $J_C(w)$ is $\{1\}$, so there are no solutions. For $J_C(5)$ the reduced Gröbner basis with respect to the lexicographic order is

$$\begin{cases} Z_5^{15} + 1, \\ Z_4^{12} + Z_4^9 Z_5^3 + Z_4^6 Z_5^6 + Z_4^3 Z_5^9 + Z_5^{12}, \\ Z_3^6 + Z_3^4 Z_4 Z_5 + Z_3^2 Z_4^2 Z_5^2 + Z_3 Z_4^4 Z_5 + Z_3 Z_4 Z_5^4 + Z_4^6 + Z_5^6, \\ g_2(Z_2, Z_3, Z_4, Z_5), \\ Z_1 + Z_2 + Z_3 + Z_4 + Z_5. \end{cases}$$

Here $g_2(Z_2, Z_3, Z_4, Z_5)$ is equal to

$$Z_2^2 + Z_2 Z_3 + Z_2 Z_4 + Z_2 Z_5 + Z_3^5 Z_4^{10} Z_5^2 + Z_3^5 Z_4^9 Z_5^3 +$$
$$+ Z_3^5 Z_4^8 Z_5^4 + Z_3^5 Z_4^7 Z_5^8 + Z_3^5 Z_3^3 Z_5^9 + Z_3^5 Z_4^2 Z_5^{10} +$$
$$+ Z_3^4 Z_4^{11} Z_5^2 + Z_3^4 Z_4^8 Z_5 + Z_3^4 Z_4^5 Z_5^8 + Z_3^4 Z_4^2 Z_5^{11} +$$
$$+ Z_3^3 Z_4^{10} Z_5^4 + Z_3^3 Z_4^9 Z_5 + Z_3^3 Z_4^8 Z_5^6 + Z_3^3 Z_4^4 Z_5^{10} +$$
$$+ Z_3^3 Z_4^3 Z_5^{11} + Z_3^3 Z_4^2 Z_5^{12} + Z_3^3 Z_5^{14} + Z_3^2 Z_4^{11} Z_5^4 +$$
$$+ Z_3^2 Z_4^8 Z_5^7 + Z_3^2 Z_4^5 Z_5^{10} + Z_3^2 Z_4^2 Z_5^{13} + Z_3^2 Z_4 Z_5^{14} +$$
$$+ Z_3^2 + Z_3 Z_4^{10} Z_5^6 + Z_3 Z_4^9 Z_5^7 + Z_3 Z_4^8 Z_5^8 + Z_3 Z_4^4 Z_5^{12} +$$
$$+ Z_3 Z_4^3 Z_5^{13} + Z_3 Z_4 + Z_4^{11} Z_5^6 + Z_4^8 Z_5^9 + Z_4^5 Z_5^{12} + Z_4^3 Z_5^{14} + Z_4^2.$$

Here already the fact that the GB of $J_C(5)$ is not equal to 1 shows that there is a solution. Theorem 1.2.7 gives all solutions explicitly. We show how to obtain one solution here. Namely, we know already that $a^{15} + 1 = 0$, where a is a primitive element of $\mathbb{F}_{16}$, so set $Z_5 = a$ and the first equation is satisfied. Substitute $Z_5 = a$ to the second equation, we have $Z_4^{12} + a^3 Z_4^9 + a^6 Z_4^6 + a^9 Z_4^3 + a^{12} = 0$. Factorizing yields that $Z_4 = 1$ is one of the roots. Substitute $Z_5 = a, Z_4 = 1$ to the third equation. We have $Z_3^6 + a Z_3^4 + a^2 Z_3^2 + Z_3 + a^{13} = 0$.

Factorizing yields that $Z_3 = a^2$ is one of the roots. Substitute $Z_5 = a, Z_4 = 1, Z_3 = a^2$ to the third equation. We have $Z_2^2 + a^{10} Z_2 + a^7 = 0$. Here $Z_2 = a^9$ is one of the roots. Finally, substitute $Z_5 = a, Z_4 = 1, Z_3 = a^2, Z_2 = a^9$ to the last equation. We obtain that $Z_1 = a^{13}$. Thus we have proved that the system $J_C(5)$ has a solution and thus the minimum distance of C is 5, which coincides with what we had in Example 1.3.2. Note that the BCH bound yields $d(C) \geq 5$, so in fact it was necessary to consider only $J_C(5)$. Here it is possible to count the number of roots. Due to the equations $Z_1^{15} - 1 = 0, \ldots, Z_5^{15} - 1 = 0$ and the fact that $\mathbb{F}_{16}$ is the splitting field of $X^{15} - 1$ we have that the number of solutions is just the product of leading terms degrees of the elements in the Gröbner basis above. This number is $15 \cdot 12 \cdot 6 \cdot 2 \cdot 1 = 2160$. Dividing this number by 5! yields the number of minimum weight codewords: 18.

We mention that the first use of Gröbner bases in finding minimum distance appears in [4].

1.3.2 Generalized Newton identities

The *error-locator polynomial* is defined by

$$\sigma(Z) = \prod_{l=1}^{t}(Z - z_l).$$

If this product is expanded

$$\sigma(Z) = Z^t + \sigma_1 Z^{t-1} + \cdots + \sigma_{t-1} Z + \sigma_t,$$

then the coefficients σ_i are the *elementary symmetric functions* in the error locations $z_1, \ldots, z_t$.

$$\sigma_i = (-1)^i \sum_{1 \leq j_1 < j_2 < \cdots < j_i \leq t} z_{j_1} z_{j_2} \ldots z_{j_i}, \ 1 \leq i \leq t.$$

Techniques for decoding cyclic codes up to half the BCH distance can be found in [108] for the binary case and [73] for arbitrary q and independently by Arimoto [3], and goes as follows. The syndromes s_i and the coefficients σ_i satisfy the following *generalized Newton identities* [108].

Theorem 1.3.5

$$s_i + \sum_{j=1}^{t} \sigma_j s_{i-j} = 0, \quad \text{for all } i \in \mathbb{Z}_n. \tag{1.7}$$

1.3. GRÖBNER BASES IN CODING THEORY: OVERVIEW

Now suppose that the complete defining set of the cyclic code contains the $2t$ consecutive elements $b, \ldots, b + 2t - 1$ for some b. Then $d \geq 2t + 1$ by the BCH bound. Furthermore the set of equations (1.7) for $i = b+t, \ldots, b + 2t - 1$ is a system of t linear equations in the unknowns $\sigma_1, \ldots, \sigma_t$ with the known syndromes $s_b, \ldots, s_{b+2t-1}$ as coefficients. Gaussian elimination solves the system of equations with complexity $\mathcal{O}(t^3)$. In this way we have obtained the *APGZ decoding* algorithm, after Arimoto, Peterson, Gorenstein and Zierler.

Example 1.3.6 We consider the same example that was considered in [7], namely of the binary 3-error correcting cyclic code of length 31 and dimension 16 with defining set $\{1, 5, 7\}$. This code is actually a quadratic residue code and has parameters [31,16,7]. The splitting field of $X^{31} - 1$ over $\mathbb{F}_2$ is $\mathbb{F}_{32}$ with a primitive 31-th root of unity a, such that $a^5 + a^2 + 1 = 0$. Note that $\mathbb{Z}_{31}$ is a disjoint union of cyclotomic classes of 1,3,5,7,11, and 15. That is to say if i is in a defining set, then $2i$ is in the complete defining set. The cyclotomic class of 1 is $\{1, 2, 4, 8, 16\}$, of 5 is $\{5, 10, 20, 9, 18\}$ and of 7 is $\{7, 14, 28, 25, 19\}$. Hence the complete defining set of C is $\{1, 2, 4, 5, 7, 8, 9, 10, 14, 16, 18, 19, 20, 25, 28\}$. It has 7,8,9,10 as four consecutive elements. Hence the BCH bound is 5 and with the APGZ algorithm we are able to correct two errors.
Let

$$\mathbf{r} = (0,0,0,0,0,1,0,0,1,0,1,1,1,1,1,0,0,1,0,0,0,1,1,1,0,1,1,0,0,0,1)$$

be a received word with at most two errors. So the known syndromes from the defining set are $s_1 = a^{13}, s_5 = a^{23}, s_7 = a^{16}$. From this we can compute $s_8 = s_1^8 = a^{11}, s_9 = s_5^8 = a^{29}, s_{10} = s_5^2 = a^{15}$. The corresponding APGZ linear system is then:

$$\begin{cases} a^{29} + a^{11}\sigma_1 + a^{16}\sigma_2 = 0, \\ a^{15} + a^{29}\sigma_1 + a^{11}\sigma_2 = 0. \end{cases}$$

This system has a unique solution $\sigma_1 = a^{13}, \sigma_2 = a^{10}$. The corresponding error locator polynomial is $\sigma(Z) = Z^2 + a^{13}Z + a^{10}$, which has the roots a^3 and a^7, so the error positions are 4 and 8. So

$$\mathbf{c} = (0,0,0,1,0,1,0,1,1,0,1,1,1,1,1,0,0,1,0,0,0,1,1,1,0,1,1,0,0,0,1)$$

is the nearest codeword.

Suppose that $\{1, \ldots, 2t\} \subseteq S_C$. Define the *syndrome polynomial* $S(Z)$ by

$$S(Z) = \sum_{i=1}^{2t} s_i Z^{i-1}.$$

The Newton identities can be reformulated as the *key equation*

$$\sigma(Z)S(Z) \equiv \omega(Z) \mod Z^{2t} \tag{1.8}$$

for some polynomial $\omega(Z)$ such that $\deg(\omega(Z)) < \deg(\sigma(Z))$. The key equation is solved by the *algorithm of Berlekamp-Massey* [15, 96] and a variant of *Euclidean algorithm* due to Sugiyama et al. [126]. Here $\omega(Z)$ is called the *error-evaluator polynomial* and is used to calculate the error values by *Forney's* formula [93], Theorem 25, p.246. Both these algorithms are of prime importance in applications. They are more efficient than solving the system of linear equations, and are basically equivalent [77], although one might prefer one over the other depending on the application and actual implementation.

All these algorithms decode up to the BCH error-correcting capacity, which is often strictly smaller than the true capacity. A general method was outlined by Berlekamp [15, pp. 231-240], Tzeng, Hartmann and Chien [128] and Stevens [124], where the *unknown syndromes* were treated as variables. We have
$$s_{i+n} = s_i, \quad \text{for all } i \in \mathbb{Z}_n,$$
since $s_{i+n} = r(a^{i+n}) = r(a^i)$. Furthermore
$$s_i^q = (e(a^i))^q = e(a^{iq}) = s_{qi}, \quad \text{for all } i \in \mathbb{Z}_n,$$
and
$$\sigma_i^{q^m} = \sigma_i, \quad \text{for all } 1 \leq i \leq t.$$
So the zeros of the following set of polynomials $Newton_t$ in the variables $S_1, \ldots, S_n$ and $\sigma_1, \ldots, \sigma_t$ is considered, see Augot et al. [5, 6].

$$Newton_t := \begin{cases} \sigma_i^{q^m} - \sigma_i, & \text{for all } 1 \leq i \leq t, \\ S_{i+n} - S_i, & \text{for all } i \in \mathbb{Z}_n, \\ S_i^q - S_{qi}, & \text{for all } i \in \mathbb{Z}_n, \\ S_i + \sum_{j=1}^{t} \sigma_j S_{i-j}, & \text{for all } i \in \mathbb{Z}_n. \end{cases} \tag{1.9}$$

It is this method of treating the unknown syndromes as variables that we generalize to arbitrary linear codes in Section 1.4.

As we have already mentioned in the introduction of Section 1.3, the solutions of $Newton_t$ are called *generic*, *formal* or *one-step* and this is considered as a preprocessing phase which has to be performed only one time. For the actual decoder for every received word $\mathbf{r}$ the variables S_i are specialized to the actual value $s_i(\mathbf{r})$ for $i \in S_C$. Alternatively one can solve $Newton_t$

1.3. GRÖBNER BASES IN CODING THEORY: OVERVIEW

together with the polynomials $S_i - s_i(\mathbf{r})$ for $i \in S_C$. This is what we called *online* decoding. Note that obtaining general error-locator polynomial as in the previous subsection is an example of formal decoding: this polynomial has to be found only once.

Example 1.3.7 Let us consider an example of decoding using Newton identities and such that the APGZ algorithm is not applicable. We consider the same 3-error correcting cyclic code of length 31 with a defining set $\{1, 5, 7\}$ as in Example 1.3.6. This time we are aiming at correcting three errors. Let us write the corresponding ideal:

$$\begin{cases} \sigma_1 S_{31} + \sigma_2 S_{30} + \sigma_3 S_{29} + S_1, \\ \sigma_1 S_1 + \sigma_2 S_{31} + \sigma_3 S_{30} + S_2, \\ \sigma_1 S_2 + \sigma_2 S_1 + \sigma_3 S_{31} + S_3, \\ \sigma_1 S_{i-1} + \sigma_2 S_{i-2} + \sigma_3 S_{i-3} + S_i, 4 \leq i \leq 31, \\ \sigma_i^{32} + \sigma_i, i = 1, 2, 3, \\ S_{i+31} + S_i, \quad \text{for all } i \in \mathbb{Z}_{31}, \\ S_i^2 + S_{2i}, \quad \text{for all } i \in \mathbb{Z}_{31}. \end{cases}$$

Note that the equations $S_{i+31} = S_i$ and $S_i^2 = S_{2i}$ imply,

$$\begin{cases} S_1^2 + S_2, & S_1^4 + S_4, & S_1^8 + S_8, & S_1^{16} + S_{16}, \\ S_3^2 + S_6, & S_3^4 + S_{12}, & S_3^8 + S_{24}, & S_3^{16} + S_{17}, \\ S_5^2 + S_{10}, & S_5^4 + S_{20}, & S_5^8 + S_9, & S_5^{16} + S_{18}, \\ S_7^2 + S_{14}, & S_7^4 + S_{28}, & S_7^8 + S_{25}, & S_7^{16} + S_{19}, \\ S_3^2 + S_6, & S_3^4 + S_{12}, & S_3^8 + S_{24}, & S_3^{16} + S_{17}, \\ S_{11}^2 + S_{22}, & S_{11}^4 + S_{13}, & S_{11}^8 + S_{26}, & S_{11}^{16} + S_{21}, \\ S_{15}^2 + S_{30}, & S_{15}^4 + S_{29}, & S_{15}^8 + S_{27}, & S_{15}^{16} + S_{23}, \\ S_{31}^2 + S_{31}. \end{cases}$$

Our intent is to write $\sigma_1, \sigma_2, \sigma_3$ in terms of known syndromes S_1, S_5, S_7. The next step would be to compute the reduced Gröbner basis of this system with respect to some elimination order induced by $S_{31} > \cdots > S_8 > S_6 > S_4 > \cdots > S_2 > \sigma_1 > \sigma_2 > \sigma_3 > S_7 > S_5 > S_1$. Unfortunately the computation is quite time consuming and the result is too huge to illustrate the idea. Rather, we do online decoding, i.e. compute syndromes S_1, S_5, S_7, plug the values into the system and then find σ's. Let

$\mathbf{r} = (0, 0, 0, 0, 0, 1, 0, 0, 1, 0, 1, 1, 1, 1, 1, 0, 0, 1, 0, 0, 0, 1, 1, 1, 0, 0, 1, 0, 0, 0, 1)$

be the received word with at most three errors. So the known syndromes we need are $s_1 = a^5, s_5 = a^8$ and $s_7 = a^{26}$. Substitute these values into the system above and compute the reduced Gröbner basis of the system. The reduced Gröbner basis with respect to the degree reverse lexicographic order (here it is possible to go without an elimination order, see Remark 1.2.8) restricted to the variables $\sigma_1, \sigma_2, \sigma_3$ is

$$\begin{cases} \sigma_3 + a^4, \\ \sigma_2 + a^5, \\ \sigma_1 + a^5. \end{cases}$$

Corresponding values for σ's give rise to the error locator polynomial:

$$\sigma(Z) = Z^3 + a^5 Z^2 + a^5 Z + a^4.$$

Factoring this polynomial yields three roots: a^3, a^7, a^{25}, which indicate error positions.

Note also that we could have worked only with the equations for $S_1, S_5, S_7, S_3, S_{11}, S_{15}, S_{31}$, but the Gröbner basis computation is harder then: on our computer it took 8 times more time.

Another way of finding the error locator polynomial $\sigma(Z)$ in the binary case is described in [7]. In this case the error values are 1 and the $S_i, i \in S_C$ are power sums of the error positions and therefore symmetric under all possible permutations of these positions. Hence S_i is equal to a polynomial $w_i(\sigma_1, \ldots, \sigma_t)$. These w_i's are known as *Waring functions*. By considering the ideal generated by the following polynomials

$$S_i - w_i(\sigma_1, \ldots, \sigma_t), i \in S_C,$$

Augot et. al. where able to prove the uniqueness theorem for the solution $(\sigma_1^*, \ldots, \sigma_t^*)$, when S_i's are assigned the concrete values of syndromes. Here the authors prefer online decoding, rather than formal one. This approach demonstrates pretty good performance in practice, but it lacks some theoretical explanations of several tricks the authors used. Further treatment of this approach is in [9].

1.3.3 Decoding affine variety codes

The method proposed by Fitzgerald and Lax [69, 70] generalizes Cooper's philosophy to arbitrary linear codes. In this approach the main notion is the *affine variety code*. Let $I = \langle g_1, \ldots, g_m \rangle \subseteq \mathbb{F}_q[X_1, \ldots, X_s]$ be an ideal. Define $I_q := I + \langle X_1^q - X_1, \ldots, X_s^q - X_s \rangle$. So I_q is a zero-dimensional ideal. Define also $V(I_q) =: \{P_1, \ldots, P_n\}$. The claim [70] is that every q-ary linear code C

1.3. GRÖBNER BASES IN CODING THEORY: OVERVIEW

with parameters $[n, k]$ can be seen as an *affine variety code* $C(I, L)$, that is the image of a vector space L of the *evaluation map*

$$\begin{cases} \phi : R \to \mathbb{F}_q^n, \\ \bar{f} \mapsto (f(P_1), \ldots, f(P_n)), \end{cases}$$

where $R := \mathbb{F}_q[U_1, \ldots, U_s]/I_q$, L is a vector subspace of R and $\bar{f}$ the coset of f in $\mathbb{F}_q[U_1, \ldots, U_s]$ modulo I_q. In order to obtain this description we do the following. Given a q-ary $[n, k]$ code C with a generator matrix $G = (g_{ij})$, we choose s, such that $q^s \geq n$, and construct s distinct points $P_1, \ldots, P_s$ in $\mathbb{F}_q^s$. Then there is an algorithm (e.g. [92]) that produces a Gröbner basis $\{g_1, \ldots, g_m\}$ for an ideal I of polynomials from $\mathbb{F}_q[X_1, \ldots, X_s]$ that vanish at the points $P_1, \ldots, P_s$. Denote by $\xi_i \in \mathbb{F}_q[X_1, \ldots, X_s]$ a polynomial that assumes the values 1 at P_i and 0 at all other P_j. The linear combinations $f_i = \sum_{i=1}^n g_{ij}\xi_j$ span the space L, so that $g_{ij} = f_i(P_j)$. In this way we obtain that the code C is the image of the evaluation above, so $C = C(I, L)$. In the same way by considering a parity check matrix instead of a generator matrix we have that the dual code is also an affine variety code.

The method of decoding is analogous to the one of CRHT with the generalization that along with the polynomials of type (1.4) one needs to add polynomials $(g_l(X_{k1}, \ldots, X_{ks}))_{l=1,\ldots,m;k=1,\ldots,t}$ for every error position. We also assume that field equations on X_{ij}'s are included among the polynomials above. Let C be a q-ary $[n, k]$ linear code such that its dual is written as an affine variety code of the form $C^\perp = C(I, L)$, where

$$\begin{cases} I = \langle g_1, \ldots, g_m \rangle \subseteq \mathbb{F}_q[X_1, \ldots, X_s], \\ L = \{\bar{f}_1, \ldots, \bar{f}_{n-k}\}, \\ V(I_q) = \{P_1, \ldots, P_s\}. \end{cases}$$

Let $\mathbf{r} = (r_1, \ldots, r_n)$ be a received word with error vector $\mathbf{e}$, so that $\mathbf{r} = \mathbf{c} + (e_1, \ldots, e_n)$ with t errors and $t \leq e$. Then the syndromes are computed by

$$s_i = \sum_{j=1}^n r_j f_i(P_j) = \sum_{j=1}^n e_j f_i(P_j) \text{ for } i = 1, \ldots, n - k.$$

Now consider the ring $\mathbb{F}_q[X_{11}, \ldots, X_{1s}, \ldots, X_{t1}, \ldots, X_{ts}, E_1, \ldots, E_t]$, where $(X_{i1}, \ldots, X_{is})$ correspond to the i-th error position and E_i to the i-th error value. Consider the ideal $\mathcal{I}_C$ generated by

$$\begin{cases} \sum_{j=1}^t E_j f_i(X_{j1}, \ldots, X_{js}) - s_i, 1 \leq i \leq n - k, \\ g_l(X_{j1}, \ldots, X_{js}), 1 \leq l \leq m, \\ E_k^{q-1} - 1. \end{cases} \quad (1.10)$$

The order $<$ is defined as follows. It is the block order $(<_1, <_2)$, where $<_1$ is the lexicographic order induced by $X_{11} < \cdots < X_{1s} < E_1$ and $<_2$ is any (e.g. degree reverse lexicographic) order on the variables $X_{21}, \ldots, X_{2s}, E_2, \ldots, X_{t1}, \ldots, X_{ts}, E_t$. We only impose lexicographic order on the first error variables, as there is a symmetry group acting on the solutions, so we are interested only in the first coordinate solutions. They will in turn give solutions for all the coordinates by symmetry. Then Theorem 2.2 from [70] states

Theorem 1.3.8 *Let G be the reduced Gröbner basis for $\mathcal{I}_C$ with respect to the order $<$. Then we may solve for the error locations and values by applying elimination theory to the polynomials in G.*

In general, finding I and L is quite technical and it turns out that for random codes this method is quite poor, because of the complicated structure of $\mathcal{I}_C$. As in Section 1.3.1 it is possible to replace syndromes with variables, but the ideal becomes then even more complicated. We consider an application of the method to Hermitian codes as is done in [70].

Example 1.3.9 Consider the Hermitian function field defined by $Y^2 + Y + X^3 = 0$ over $\mathbb{F}_4$ with a primitive element a, such that $a^2 + a + 1 = 0$. Let C be a quaternary [8,3,5] Hermitian code. It is orthogonal to the [8,5,3] Hermitian code defined by $L = \langle 1, X, Y, X^2, XY \rangle$. Choose $s = 2$, so that $4^2 > 8$, and the points

$$P_1 = (0,0), \quad P_2 = (0,1), \quad P_3 = (1,a), \quad P_4 = (1,a^2),$$
$$P_5 = (a,a), \quad P_6 = (a,a^2), \quad P_7 = (a^2,a), \quad P_8 = (a^2,a^2).$$

Denote $I = \langle Y^2 + Y + X^3 \rangle$. Then $C^\perp = C(I, L)$. The parity check matrix for C is

$$H = \begin{pmatrix} 1 & 1 & 1 & 1 & 1 & 1 & 1 & 1 \\ 0 & 0 & 1 & 1 & a & a & a^2 & a^2 \\ 0 & 1 & a & a^2 & a & a^2 & a & a^2 \\ 0 & 0 & 1 & 1 & a^2 & a^2 & a & a \\ 0 & 0 & a & a^2 & a^2 & 1 & 1 & a \end{pmatrix}.$$

Let $\mathbf{r} = (1, 1, 0, a, a^2, a^2, a, 1)$ be a received word with at most two errors. The corresponding syndrome is $\mathbf{s} = (1, 0, a^2, a^2, 0)$. So the ideal $\mathcal{I}_C$ in

1.3. GRÖBNER BASES IN CODING THEORY: OVERVIEW

$\mathbb{F}_4[X_1, Y_1, E_1, X_2, Y_2, E_2]$ is generated by

$$\begin{cases} X_1^4 + X_1, & X_2^4 + X_2, \\ Y_1^4 + Y_1, & Y_2^4 + Y_2, \\ E_1^3 + 1, & E_2^3 + 1, \\ Y_1^2 + Y_1 + X_1^3, & Y_2^2 + Y_2 + X_2^3, \\ E_1 + E_2 + 1, \\ E_1 X_1 + E_2 X_2, \\ E_1 Y_1 + E_2 Y_2 + a^2, \\ E_1 X_1^2 + E_2 X_2^2 + a^2, \\ E_1 X_1 Y_1 + E_2 X_2 Y_2. \end{cases}$$

We are working with a block order $(<_1, <_2)$ induced by $X_1 < Y_1 < E_1 < X_2 < Y_2 < E_2$, where $<_1$ is the lexicographic order induced by $X_1 < Y_1 < E_1$, and $<_2$ is the degree reverse lexicographic order induced by $X_2 < Y_2 < E_2$. The reduced Gröbner basis G of $\mathcal{I}_C$ with respect to this order is

$$\begin{cases} X_1^2 + aX_1 + a^2, \\ Y_1 + a^2, \\ E_1 + a^2 X_1 + 1, \\ X_2 + X_1 + a, \\ Y_2 + a^2, \\ E_2 + a^2 X_1 \end{cases}$$

Solving the first equation $X_1^2 + aX_1 + a^2 = 0$ actually gives the X-coordinates of the two error positions. They are 1 and a^2. We substitute further and obtain the error positions $(1, a^2)$ and (a^2, a^2), that is positions 4 and 8 in our numeration, and the corresponding error values a and a^2, respectively. Hence $\mathbf{c} = (1, 1, 0, 0, a^2, a^2, a, a)$ is the codeword sent.

We mention that there are generalizations of the approach of Fitzgerald and Lax, which follow the same idea as the generalizations for the CRHT-ideal. Namely, one adds the polynomials that ensure that the error locations are different. For more details, see [115]. There it is also proven that affine variety codes possess the so-called *multi-dimensional general error-locator polynomial*, which is a generalization of the general error locator polynomial from Section 1.3.1.

1.3.4 Syndrome decoding with Gröbner bases

In this section we give a formulation of the well-known syndrome decoding in terms of ideals and solutions of the corresponding systems. Moreover, some results of this section (e.g. Lemma 1.3.18) are later used in subsection 1.4.2, where we look closely at the structure of ideals that we need in our construction for decoding.

Choose a parity check matrix H of an $[n, k, d]$ code C. Let $\mathbf{h}_1, \ldots, \mathbf{h}_{n-k}$ be the rows of H.

Remark 1.3.10 Let $\tilde{C} = \mathbb{F}_{q^m} C$ be the code over $\mathbb{F}_{q^m}$ for some m that is generated by C. Then C is the restriction of $\tilde{C}$ to $\mathbb{F}_q^n$, that is $C = \mathbb{F}_q^n \cap \tilde{C}$. And H is also a parity check matrix of $\tilde{C}$, since the rank of H does not change under the extension from $\mathbb{F}_q$ to $\mathbb{F}_{q^m}$. Furthermore C and $\tilde{C}$ have the same minimum distance, since it is equal to the minimum number of dependent columns of H, and this does not change under an extension of scalars.

Definition 1.3.11 The *(known) syndrome* $\mathbf{s}(H, \mathbf{y})$ of a word $\mathbf{y}$ with respect to H is the column vector $\mathbf{s}(H, \mathbf{y}) = H\mathbf{y}^T$. It has entries $s_i(H, \mathbf{y}) = \mathbf{h}_i \cdot \mathbf{y}$ for $i = 1, \ldots, n-k$, where $\mathbf{h}_i \cdot \mathbf{y}$ is inner product of $\mathbf{h}_i$ and $\mathbf{y}$. The abbreviations $\mathbf{s}(\mathbf{y})$ and $s_i(\mathbf{y})$ are used for $\mathbf{s}(H, \mathbf{y})$ and $s_i(H, \mathbf{y})$, respectively.

Remark 1.3.12 Let $\mathbf{r} = \mathbf{c} + \mathbf{e}$ be a *received word* with $\mathbf{c} \in C$ the codeword that was sent and $\mathbf{e}$ the *error vector*. Then $\mathbf{h}_i \cdot \mathbf{c} = 0$ for all $i = 1, \ldots, n-k$. So the syndromes of $\mathbf{r}$ and $\mathbf{e}$ with respect to H are equal and known:

$$s_i(\mathbf{r}) := \mathbf{h}_i \cdot \mathbf{r} = \mathbf{h}_i \cdot \mathbf{e} = s_i(\mathbf{e}).$$

Let $\mathbf{h}'_1, \ldots, \mathbf{h}'_n$ be the n columns of H. If furthermore the support of $\mathbf{e}$ is equal to $\{i_1, \ldots, i_t\}$, then

$$\mathbf{s}(\mathbf{r}) = \mathbf{s}(\mathbf{e}) = e_{i_1} \mathbf{h}'_{i_1} + \cdots + e_{i_t} \mathbf{h}'_{i_t}.$$

Therefore, if the distance of a received word to the code is t, then the syndrome vector of the received word is a linear combination of t columns of H. By *syndrome decoding* here we mean an algorithm that finds such a linear combination. One way to accomplish this is to go though all possible t-subsets of $\{1, \ldots, n\}$ and see by linear algebra whether a linear combination of the corresponding columns of H gives the syndrome vector. The complexity is therefore $\mathcal{O}(\binom{n}{t}(n-k)t^2)$.

Finding the minimum distance is similar, since we take the syndrome equal to the zero vector, so we try to find the smallest number of columns of H that are linearly dependent.

Definition 1.3.13 Let $\mathbf{r} \in \mathbb{F}_q^n$ and let $d(\mathbf{r}, C)$ be the distance of $\mathbf{r}$ to C. A *nearest codeword* of $\mathbf{r}$ to C is an element $\mathbf{c} \in C$ such that $d(\mathbf{r}, \mathbf{c}) = d(\mathbf{r}, C)$. Let $\mathcal{L}(\mathbf{r}, C)$ be the list of nearest codewords of $\mathbf{r}$ to C.

1.3. GRÖBNER BASES IN CODING THEORY: OVERVIEW

Proposition 1.3.14 *Let $\tilde{C} = \mathbb{F}_{q^m} C$. If $\mathbf{r} \in \mathbb{F}_q^n$, then $d(\mathbf{r}, C) = d(\mathbf{r}, \tilde{C})$ and $\mathcal{L}(\mathbf{r}, C) = \mathcal{L}(\mathbf{r}, \tilde{C})$.*

Proof. (1) We have $d(\mathbf{r}, C) \geq d(\mathbf{r}, \tilde{C})$, since $C \subseteq \tilde{C}$. There are $d(\mathbf{r}, \tilde{C})$ columns of H such that an $\mathbb{F}_{q^m}$-linear combination of these columns is equal to $\mathbf{s}(H, \mathbf{r})$. But $\mathbf{r}$ and H have entries in $\mathbb{F}_q$. Hence $d(\mathbf{r}, C) \leq d(\mathbf{r}, \tilde{C})$. Therefore equality holds.
(2) We have $\mathcal{L}(\mathbf{r}, C) \subseteq \mathcal{L}(\mathbf{r}, \tilde{C})$ by (1). Conversely, let $\mathbf{c} \in \mathcal{L}(\mathbf{r}, \tilde{C})$ and $t = d(\mathbf{r}, \tilde{C})$. Let $\mathbf{e} = \mathbf{r} - \mathbf{c}$. Let $I = \{i_1, \ldots, i_t\}$ be the support of $\mathbf{e}$ that is the set of nonzero coordinates of $\mathbf{e}$. Let H_I be the submatrix of H consisting of the columns $h_{i_1}, \ldots, h_{i_t}$. Let $\mathbf{s} = H\mathbf{r}^T$. Then $\mathbf{s}$ is a linear combination of the columns of H_I. So H_I and the extended matrix $[H_I|\mathbf{s}]$ have the same rank. This rank is t, otherwise we would have a proper subset I' of I such that $H_{I'}$ and H_I have the same rank. But this would give an $\mathbf{e}'$ with support I' of weight $t' < t$ and $H\mathbf{e}'^T = \mathbf{s}$. This gives $\mathbf{c}' \in \tilde{C}$ with $\mathbf{r} = \mathbf{c}' + \mathbf{e}'$. So $d(\mathbf{r}, \tilde{C}) \leq t' < t$, a contraction. Hence H_I and the extended matrix $[H_I|\mathbf{s}]$ have the same rank t. So $H_I \mathbf{x}^T = \mathbf{s}$ has a unique solution $\mathbf{x} = (e_{i_1}, \ldots, e_{i_t})$ with entries in $\mathbb{F}_q$. Hence and $\mathbf{c} = \mathbf{r} - \mathbf{e} \in \mathbb{F}_q^n \cap \tilde{C} = C$. Therefore $\mathbf{c} \in \mathcal{L}(\mathbf{r}, C)$. ◊

Definition 1.3.15 *Let $h_i(E)$ be the linear function in $\mathbb{F}_q[E_1, \ldots, E_n]$ defined by*
$$h_i(E) = \sum_{j=1}^n h_{ij} E_j,$$
where $H = (h_{ij})$. Let $E(\mathbf{r})$ be the ideal in $\mathbb{F}_q[E_1, \ldots, E_n]$ generated by the elements $h_i(E) - s_i(\mathbf{r})$ for all $i = 1, \ldots, n - k$. Let $J(t, n)$ be the ideal in $\mathbb{F}_q[E_1, \ldots, E_n]$ defined by
$$J(t, n) = \bigcap_{1 \leq j_1 < \cdots < j_{n-t} \leq n} \langle E_{j_1}, \ldots, E_{j_{n-t}} \rangle.$$

Let $E(t, \mathbf{r})$ be the ideal generated by $E(\mathbf{r})$ and $J(t, n)$.

Lemma 1.3.16
1) $\mathbf{e}$ is a solution of $E(\mathbf{r})$ if and only if $\mathbf{r} = \mathbf{c} + \mathbf{e}$ for $\mathbf{c} \in \mathbb{F}_{q^m} C$ for some positive integer m.
2) $\mathbf{e}$ is a solution of $J(t, n)$ if and only if $wt(\mathbf{e}) \leq t$.
3) Let $t = d(\mathbf{r}, C)$. Then $E(w, \mathbf{r})$ has no solution for all $w < t$. And $\mathbf{e}$ is a solution of $E(t, \mathbf{r})$ if and only if $\mathbf{r} = \mathbf{c} + \mathbf{e}$ for some $\mathbf{c} \in C$ and $wt(\mathbf{e}) = t$.

Proof.
1) If $\mathbf{r} = \mathbf{c} + \mathbf{e}$ for some $\mathbf{c} \in \mathbb{F}_{q^m}C$, then $\mathbf{e}$ is a solution of $E(\mathbf{r})$ by Remarks 1.3.10 and 1.3.12.
Conversely, if $\mathbf{e}$ is a solution of $E(\mathbf{r})$, then $\mathbf{s}(\mathbf{r}) = \mathbf{s}(\mathbf{e})$ and $\mathbf{e} \in \mathbb{F}_{q^m}^n$ for some positive integer m. So $\mathbf{s}(\mathbf{r} - \mathbf{e}) = 0$. Hence $\mathbf{c} = \mathbf{r} - \mathbf{e}$ is a codeword of $\mathbb{F}_{q^m}C$ and $\mathbf{r} = \mathbf{c} + \mathbf{e}$.
2) If $wt(\mathbf{e}) \leq t$, then the support of $\mathbf{e}$ is contained in $\{k_1, \ldots, k_t\}$ for some $\mathbf{k}$. Let $\{j_1, \ldots, j_{n-t}\}$ be the complement of this support. Then $\mathbf{e}$ is a solution of the ideal $\langle E_{j_1}, \ldots, E_{j_{n-t}} \rangle$. So $\mathbf{e}$ is an element of the variety $V(\langle E_{j_1}, \ldots, E_{j_{n-t}} \rangle)$. Hence $\mathbf{e}$ is an element of the variety

$$\bigcup_{1 \leq j_1 < \cdots < j_{n-t} \leq n} V(\langle E_{j_1}, \ldots, E_{j_{n-t}} \rangle) =$$

$$V\left(\cap_{1 \leq j_1 < \cdots < j_{n-t} \leq n} \langle E_{j_1}, \ldots, E_{j_{n-t}} \rangle\right) = V(J(t, n)).$$

So $\mathbf{e}$ is a solution of $J(t, n)$
The converse is proved similarly.
3) Is a direct consequence of (1) and (2) and Proposition 1.3.14. ◇

Theorem 1.3.17 *Let H be a parity check matrix of the code C. Let $\mathbf{r}$ be a received word. Let t be the smallest positive integer such that $E(t, \mathbf{r})$ has a solution.*
1) Then the solutions $\mathbf{e}$ of $E(t, \mathbf{r})$ correspond one-to-one to $\mathbf{c}$ in $\mathcal{L}(\mathbf{r}, C)$.
2) If $\mathbf{e}$ is a solution of $E(t, \mathbf{r})$ such that $wt(\mathbf{e}) \leq (d(C) - 1)/2$, then $wt(\mathbf{e}) = t$ and $\mathbf{e}$ is the unique solution.

Proof. (1) is a direct consequence of Lemma 1.3.16 (3).
(2) is a consequence of (1) and the well-known fact that $\mathbf{r}$ has a unique nearest codeword in case $d(\mathbf{r}, C) \leq (d - 1)/2$. ◇

Lemma 1.3.18 *Let*

$$I(t, n) = \langle E_{i_1} \cdots E_{i_{t+1}} | 1 \leq i_1 < \cdots < i_{t+1} \leq n \rangle.$$

Then $I(t, n) = J(t, n)$.

Proof. Let $E_{i_1} \cdots E_{i_{t+1}}$ be a generator of the ideal $I(t, n)$ and let $\mathbf{j} = \{j_1, \ldots, j_{n-t}\}$ be an increasing $n - t$ tuple. Then $\{i_1, \ldots, i_{t+1}\}$ and $\{j_1, \ldots, j_{n-t}\}$ are subsets of $\{1, \ldots, n\}$ consisting of $t + 1$ and $n - t$ elements, respectively. Hence their intersection is not empty, that is $i_r = j_s$ for some r and s. Hence

$$E_{i_1} \cdots E_{i_{t+1}} \in \langle E_{j_s} \rangle \subseteq \langle E_{j_1}, \ldots, E_{j_{n-t}} \rangle.$$

1.4. METHODS BASED ON QUADRATIC EQUATIONS

Thus it is proved that $I(t,n) \subseteq J(t,n)$.

Now consider $J(t,n)/I(t,n)$. Let f be a polynomial in $J(t,n)$. Modulo $I(t,n)$ we may assume that

$$f = \sum_{\mathbf{i}=(i_1,\ldots,i_t), 1 \leq i_1 < \cdots < i_t \leq n} f_{\mathbf{i}} E_{i_1}^{\alpha_{i_1}} \cdots E_{i_t}^{\alpha_{i_t}}$$

with $f_{\mathbf{i}} \in \mathbb{F}_q$ and α_i are non-negative integers, which depend on $\mathbf{i}$. For every $\mathbf{i}$ with $1 \leq i_1 < \cdots < i_t \leq n$ there exists exactly one $\mathbf{j}$ such that $1 \leq j_1 < \cdots < j_{n-t} \leq n$ and $\{j_1, \ldots, j_{n-t}\}$ is the complement of $\{i_1, \ldots, i_t\}$ in $\{1, \ldots, n\}$. Now $f \in J(t,n)$. So $f \in \langle E_{j_1}, \ldots, E_{j_{n-t}} \rangle$. This is only possible if $f_{\mathbf{i}} = 0$ since the sets of variables $\{E_{i_1}, \ldots, E_{i_t}\}$ and $\{E_{j_1}, \ldots, E_{j_{n-t}}\}$ are disjoint. So $f_{\mathbf{i}} = 0$ for every such $\mathbf{i}$. Hence $f = 0$. Therefore $J(t,n) \subseteq I(t,n)$. ◊

Remark 1.3.19 The ideal $J(t,n)$ is generated by $\binom{n}{t+1}$ monomials of degree $t+1$, and thus $E(t, \mathbf{r})$ is generated by $n-k$ linear functions and $\binom{n}{t+1}$ monomials of degree $t+1$.

Example 1.3.20 We provide now a small example explaining Theorem 1.3.17. Consider a one error-correcting Hamming code with parameters $[7,4,3]$ over $\mathbb{F}_2$. Let $\mathbf{r} = (1,0,1,0,1,1,1)$ be a received word. Let a parity check matrix be

$$H = \begin{pmatrix} 1 & 1 & 0 & 1 & 1 & 0 & 0 \\ 1 & 0 & 1 & 1 & 0 & 1 & 0 \\ 1 & 1 & 1 & 0 & 0 & 0 & 1 \end{pmatrix}.$$

The ideal $E(1, \mathbf{r})$ is generated by the linear polynomials $E_1+E_2+E_4+E_5$, $E_1+E_3+E_4+E_6+1$, $E_1+E_2+E_3+E_7+1$ and the binomials $E_i E_j$, $1 \leq i < j \leq 7$. Now $E(1, \mathbf{r})$ has $(0,0,1,0,0,0,0)$ as the unique solution in the algebraic closure. Moreover, the same situation takes place when we consider e.g. a code with the parity check matrix H over $\mathbb{F}_8$. In particular, one does not need to add field equations in order to avoid *spurious* solutions, i.e. solutions that do not correspond to codeword(s).

1.4 Methods based on quadratic equations

In the previous sections we have discussed existing methods for decoding and finding minimum distance of cyclic codes. The method for decoding arbitrary linear codes due to Fitzgerald and Lax quite obviously leads to very complicated polynomial systems when one works with codes with no specific structure or if one wants to correct more, than just a few errors, see Section 1.4.8. The aim of the following exposition is to present a method which would be applicable to arbitrary linear codes and had more inner structure. The polynomial quadratic system will be obtained that in certain setting generalizes the idea from Section 1.3.2. The peculiar feature of our approach is that despite the fact that we want to obtain solutions in some finite field, in our method it is not necessary to add field equations, cf. Section 1.3.4. Preparation material is given in Sections 1.4.1 and 1.4.2, the decoding method itself is further in Section 1.4.3. In Section 1.4.4 we talk about how our method generalizes Newton identities method. Section 1.4.5 is devoted to finding the minimum distance and inherits the approach developed before. We conclude with some discussions on complexity issues in Section 1.4.7 and experimental material in Section 1.4.8.

1.4.1 Matrix in MDS form

In this section we introduce the notions of an MDS basis/matrix and an unknown syndrome. We establish some properties of the matrix of unknown syndromes.

Let $\mathbf{b}_1, \ldots, \mathbf{b}_n$ be a basis of $\mathbb{F}_q^n$. Next, let B is the $n \times n$ matrix with $\mathbf{b}_1, \ldots, \mathbf{b}_n$ as rows.

Definition 1.4.1 The *(unknown) syndrome* $\mathbf{u}(B, \mathbf{e})$ of a word $\mathbf{e}$ with respect to B is the column vector $\mathbf{u}(B, \mathbf{e}) = B\mathbf{e}^T$. It has entries $u_i(B, \mathbf{e}) = \mathbf{b}_i \cdot \mathbf{e}$ for $i = 1, \ldots, n$.

Remark 1.4.2 The matrix B is invertible, since its rank is n. The syndrome $\mathbf{u}(B, \mathbf{e})$ determines the (error) vector $\mathbf{e}$ uniquely, since

$$B^{-1}\mathbf{u}(B, \mathbf{e}) = B^{-1}B\mathbf{e}^T = \mathbf{e}^T.$$

So the idea is based on finding the unknown syndrome of an error vector $\mathbf{e}$ with respect to some specific basis B. Then finding the error vector is trivial. From now on $\mathbf{u}(B, \mathbf{e})$ is abbreviated by $\mathbf{u}(\mathbf{e})$, and $\mathbf{u}^T B^{-T}$ is denoted by $\mathbf{e}(\mathbf{u})$. We have a linear automorphism β of $\overline{\mathbb{F}_q}^n$, defined by $\beta(\mathbf{e}) = \mathbf{e}B^T$ with inverse map γ. This induces an isomorphism of rings

$$\beta^* : \mathbb{F}_q[U_1, \ldots, U_n] \longrightarrow \mathbb{F}_q[E_1, \ldots, E_n],$$

defined by

$$\beta^*(U_i) = \sum_{j=1}^{n} b_{ij} E_j$$

1.4. METHODS BASED ON QUADRATIC EQUATIONS

and its inverse γ^*, defined by

$$\gamma^*(E_i) = \sum_{j=1}^{n} c_{ij} U_j,$$

where the c_{ij} are the entries of B^{-1}.

Definition 1.4.3 Define the coordinatewise *star product* of two vectors $\mathbf{x}, \mathbf{y} \in \mathbb{F}_q^n$ by $\mathbf{x} * \mathbf{y} = (x_1 y_1, \ldots, x_n y_n)$. Then $\mathbf{b}_i * \mathbf{b}_j$ is a linear combination of the basis vectors $\mathbf{b}_1, \ldots, \mathbf{b}_n$, i.e. there are constants $\mu_l^{ij} \in \mathbb{F}$ such that

$$\mathbf{b}_i * \mathbf{b}_j = \sum_{l=1}^{n} \mu_l^{ij} \mathbf{b}_l.$$

The elements $\mu_l^{ij} \in \mathbb{F}_q$ are called the *structure constants* of the basis $\mathbf{b}_1, \ldots, \mathbf{b}_n$, see [84].

Definition 1.4.4 Define the $n \times n$ matrix of *(unknown) syndromes* $\mathcal{U}(\mathbf{e})$ of a word $\mathbf{e}$ by $u_{ij}(\mathbf{e}) = (\mathbf{b}_i * \mathbf{b}_j) \cdot \mathbf{e}$. The following abbreviations $\mathbf{u}(\mathbf{e})$ and $u_i(\mathbf{e})$ are used for $\mathbf{u}(B, \mathbf{e})$ and $u_i(B, \mathbf{e})$, respectively.

Remark 1.4.5 The relation between the entries of the matrix $\mathcal{U}(\mathbf{e})$ and the vector $\mathbf{u}(\mathbf{e})$ of unknown syndromes is given by

$$u_{ij}(\mathbf{e}) = \sum_{l=1}^{n} \mu_l^{ij} u_l(\mathbf{e}).$$

Proposition 1.4.6 *Let $D(\mathbf{e})$ be the diagonal matrix with $\mathbf{e}$ on its diagonal. Then*

$$\mathcal{U}(\mathbf{e}) = B D(\mathbf{e}) B^T,$$

and the rank of $\mathcal{U}(\mathbf{e})$ is equal to the weight of $\mathbf{e}$.

Proof. See also [79, Lemma 4.7]. We have that

$$u_{ij}(\mathbf{e}) = (\mathbf{b}_i * \mathbf{b}_j) \cdot \mathbf{e} = \sum_{l=1}^{n} b_{il} e_l b_{jl}.$$

Hence $\mathcal{U}(\mathbf{e}) = B D(\mathbf{e}) B^T$. Now B an B^T are invertible. Hence the rank of $\mathcal{U}(\mathbf{e})$ is equal to the rank of $D(\mathbf{e})$ which is equal to the weight of $\mathbf{e}$. ◇

So there are $\text{wt}(\mathbf{e}) + 1$ columns of $\mathcal{U}(\mathbf{e})$ that are dependent and every w tuple of columns of $\mathcal{U}(\mathbf{e})$ are independent if $w \leq \text{wt}(\mathbf{e})$. We will look at the smallest t such that the first $t + 1$ columns are dependent. Consider the

$v \times w$ matrix $\mathcal{U}_{vw}(\mathbf{y})$ of a word $\mathbf{y}$ with the entries $u_{ij}(\mathbf{y})$ for $i = 1, \ldots, v$ and $j, \ldots, w$. For an arbitrary matrix B we have to go through all the $\binom{n}{w}$ w-tuples of columns of B with $w \leq \mathrm{wt}(\mathbf{e}) + 1$ to find such a dependency. This is not efficient. There is a more efficient way with the help of a B in special form.

Definition 1.4.7 Let $\mathbf{b}_1, \ldots, \mathbf{b}_n$ be a basis of $\mathbb{F}_q^n$. Let B_s be the $s \times n$ matrix with $\mathbf{b}_1, \ldots, \mathbf{b}_s$ as rows, so $B = B_n$. We say that $\mathbf{b}_1, \ldots, \mathbf{b}_n$ is an ordered *MDS basis* and B an *MDS matrix* if all the $s \times s$ submatrices of B_s have rank s for all $s = 1, \ldots, n$.

Remark 1.4.8 Let B be an MDS matrix. Let C_s be the code with B_s as parity check matrix. Then C_s is an MDS (Maximum Distance Separable) code for all s. This motivates the name in the previous definition.

Definition 1.4.9 Suppose $n \leq q$. Let $\mathbf{x} = (x_1, \ldots, x_n)$ be an n-tuple of pairwise distinct elements in $\mathbb{F}_q$. Define

$$\mathbf{b}_i = (x_1^{i-1}, \ldots, x_n^{i-1}).$$

Then $\mathbf{b}_1, \ldots, \mathbf{b}_n$ is called an ordered *Vandermonde basis* and the corresponding matrix is denoted by $B(\mathbf{x})$ and called a *Vandermonde matrix*.
In particular, if $\alpha \in \mathbb{F}_q^*$ is an element of order n and $x_j = \alpha^{j-1}$ for all j, then $\mathbf{b}_1, \ldots, \mathbf{b}_n$ is called an ordered *Reed-Solomon (RS) basis* and the corresponding matrix is called a *RS matrix* and denoted by $B(\alpha)$. Sometimes we will refer to an *MDS extension* of some field $\mathbb{F}_q$ as to a field extension $\mathbb{F}_{q^m}$, such that an MDS basis exists for $\mathbb{F}_{q^m}^n$.

Remark 1.4.10 If $\mathbf{b}_1, \ldots, \mathbf{b}_n$ is a Vandermonde basis of $\mathbb{F}_q^n$, then it is an MDS basis.
For a finite field and general n there is a positive integer m such that $n \leq q^m$. The above construction gives a Vandermonde basis $\mathbf{b}_1, \ldots, \mathbf{b}_n$ of $\mathbb{F}_{q^m}^n$ over $\mathbb{F}_{q^m}$ such that

$$\mathbf{b}_i * \mathbf{b}_j = \mathbf{b}_{i+j-1} \quad \text{and} \quad u_{ij}(\mathbf{e}) = u_{i+j-1}(\mathbf{e}) \quad \text{if} \quad i+j \leq n+1.$$

In case n and q are relatively prime, n divides $q^m - 1$ for some m, and we can take an element $\alpha \in \mathbb{F}_{q^m}^*$ of order n and $x_j = \alpha^{j-1}$ for $i = 1, \ldots, n$ like for cyclic codes. In that case $u_{ij}(\mathbf{e}) = u_{((i+j-2) \mod n)+1}(\mathbf{e})$ for all $1 \leq i, j \leq n$.

1.4. METHODS BASED ON QUADRATIC EQUATIONS

Remark 1.4.11 For any prime p and positive integer M there is an algorithm of polynomial computing time $(p \log M)^{\mathcal{O}(1)}$ that computes an irreducible polynomial of degree $m = M + o(M)$ over $\mathbb{F}_p$, see [118, 119]. Hence for a given field $\mathbb{F}_q$, the complexity of finding an extension $\mathbb{F}_{q^m}$ such that $q^m \geq n$, is polynomial in n.

Definition 1.4.12 Let M be a matrix with an entry m_{ij} in row i and column j. Then M_v is the submatrix of M consisting of the first v columns, and M_{uv} is the $u \times v$ submatrix of M given by

$$M_{uv} = \begin{pmatrix} m_{11} & m_{12} & \cdots & m_{1v} \\ m_{21} & m_{22} & \cdots & m_{2v} \\ \vdots & \vdots & \ddots & \vdots \\ m_{u1} & m_{u2} & \cdots & m_{uv} \end{pmatrix}.$$

Proposition 1.4.13 *Suppose that B is an MDS matrix. Let $w = \mathrm{wt}(\mathbf{e})$. Then*

$$\mathrm{rank}(\mathcal{U}_{nv}(\mathbf{e})) = \min\{v, w\}.$$

Proof. We have that

$$u_{ij}(\mathbf{e}) = (\mathbf{b}_i \ast \mathbf{b}_j) \cdot \mathbf{e} = \sum_{l=1}^n b_{il} e_l b_{jl}.$$

Hence

$$\mathcal{U}_{nv}(\mathbf{e}) = B D(\mathbf{e}) B_v^T,$$

where $D(\mathbf{e})$ is the diagonal matrix with $\mathbf{e}$ on the diagonal. This triple product implies that $\mathrm{rank}(\mathcal{U}_{nv}(\mathbf{e})) \leq \min\{v, w\}$, since $\mathrm{rank}(B) = n$, $\mathrm{rank}(B_v) = v$ and $\mathrm{rank}(D(\mathbf{e})) = w$.
We may assume without loss of generality that the nonzero entries of $\mathbf{e}$ are at the beginning, since B stays an MDS matrix after a permutation of its columns. So the non-zero entries are $e_1, \ldots, e_w$. Let $\mathbf{e}' = (e_1, \ldots, e_w)$. Then $BD(\mathbf{e})B_v^T$ has $B_w D(\mathbf{e}')B_{vw}^T$ as a submatrix. Now $D(\mathbf{e}')$ is invertible, since all the coordinates of $\mathbf{e}'$ are nonzero. Hence $B_w D(\mathbf{e})B_v^T$ has the same rank as $B_w B_{vw}^T$. Now B_{ww} is invertible, since B is an MDS matrix. Hence $B_w B_{vw}^T$ has the same rank as B_{vw}^T. But $\mathrm{rank}(B_{vw}) = \min\{v, w\}$, again since B is MDS. Therefore $\mathrm{rank}(\mathcal{U}_{nv}(\mathbf{e})) \geq \min\{v, w\}$. ◇

Example 1.4.14 Let q be a power of a prime p, $p \neq 3$. Let $\alpha \in \mathbb{F}_q$ such that $1 + \alpha + \alpha^2 = 0$. Then the order of α is 3. Take the matrix B over $\mathbb{F}_q$ given by

$$B = \begin{pmatrix} 1 & 1 & 1 \\ 1 & \alpha & \alpha^2 \\ 1 & \alpha^2 & \alpha \end{pmatrix}.$$

Then B is in MDS form. Take $\mathbf{e} = (1,1,1)$. Then

$$\mathcal{U}_{33}(\mathbf{e}) = \begin{pmatrix} 3 & 0 & 0 \\ 0 & 0 & 3 \\ 0 & 3 & 0 \end{pmatrix}$$

has rank 3. But $\mathcal{U}_{22}(\mathbf{e})$ has rank 1. So the following strengthening of Lemma 1.4.13 for an MDS matrix B with $w = \text{wt}(\mathbf{e})$:

$$\text{rank}(\mathcal{U}_{uv}(\mathbf{e})) = \min\{u, v, w\} \text{ is not true.}$$

1.4.2 Determinantal variety of syndromes

In the previous subsection it is shown that $\mathcal{U}_{nv}(\mathbf{e})$ has rank v if $v \leq \text{wt}(\mathbf{e})$, and its rank is $\text{wt}(\mathbf{e})$ if $v > \text{wt}(\mathbf{e})$. We see that the first moment of stabilization of the rank of $\mathcal{U}_{nv}(\mathbf{e})$ yields the weight of $\mathbf{e}$. We would like to be able to find this moment. For this we change all the entries of $\mathcal{U}_{nv}(\mathbf{e})$ to variables and search for linear dependence of columns, see Definitions 1.4.15 and 1.4.29. The corresponding results are developed in this section. It turns out that the study of the corresponding ideal also gives an opportunity to find a vector of unknown syndromes $\mathbf{u}$.

Definition 1.4.15 Let B be an MDS matrix with structure constants μ_l^{ij}. Define the linear functions U_{ij} in the variables $U_1, \ldots, U_n$ by

$$U_{ij} = \sum_{l=1}^{n} \mu_l^{ij} U_l.$$

Let $\mathcal{U}$ be the $n \times n$ matrix with entries U_{ij}.

Remark 1.4.16 If $U_i = u_i(\mathbf{e})$ for all i, then $U_{ij} = u_{ij}(\mathbf{e})$ for all i, j. So the matrix above exactly reflects the idea stated at the beginning of this subsection.

1.4. METHODS BASED ON QUADRATIC EQUATIONS

Definition 1.4.17 Let $\mathbf{b}'_1, \ldots, \mathbf{b}'_n$ be the columns of B. Let $\mathbf{i} = (i_1, \ldots, i_t)$ with $1 \leq i_1 < \cdots < i_t \leq n$. Let $L(\mathbf{i})$ be the linear subspace of the vector space $\overline{\mathbb{F}_q}^n$ generated by $\mathbf{b}'_{i_1}, \ldots, \mathbf{b}'_{i_t}$. That is

$$L(\mathbf{i}) = \{\, x_1 \mathbf{b}'_{i_1} + \cdots + x_t \mathbf{b}'_{i_t} \mid x_1, \ldots, x_t \in \overline{\mathbb{F}_q} \,\}.$$

Let $I(\mathbf{i})$ be the defining ideal of $L(\mathbf{i})$ in $\mathbb{F}_q[U_1, \ldots, U_n]$. That is

$$I(\mathbf{i}) = \{\, f(U) \in \mathbb{F}_q[U_1, \ldots, U_n] \mid f(\mathbf{u}) = 0 \text{ for all } \mathbf{u} \in L(\mathbf{i}) \,\}.$$

Lemma 1.4.18 *Let $\mathbf{k}$ be a t-tuple of increasing entries and let $\{j_1, \ldots, j_{n-t}\}$ be the complement of $\{k_1, \ldots, k_t\}$ in $\{1, \ldots, n\}$. Then $L(\mathbf{k})$ has dimension t and the ideal $I(\mathbf{k})$ is a radical ideal generated by the $n-t$ linear functions $\gamma^*(E_{j_1}), \ldots, \gamma^*(E_{j_{n-t}})$ in the variables $U_1, \ldots, U_n$.*

Proof. The dimension of $L(\mathbf{k})$ follows form Definition 1.4.17 and the fact that $\mathbf{b}'_1, \ldots, \mathbf{b}'_n$ are independent. Now $\mathbf{u} = \mathbf{u}(\mathbf{e}) = B\mathbf{e}^T = e_1 \mathbf{b}'_1 + \cdots + e_n \mathbf{b}'_n$. Furthermore $\mathrm{wt}(\mathbf{e}) \leq t$ if and only if the support of $\mathbf{e}$ is contained in $\{k_1, \ldots, k_t\}$ for some $\mathbf{k}$ with $1 \leq k_1 < \cdots < k_t \leq n$. Let β and γ be the maps of Remark 1.4.2. Then $\mathbf{u} \in I(\mathbf{k})$ if and only if $\mathbf{e} = \gamma(\mathbf{u})$ and $e_{j_1} = \cdots = e_{j_{n-t}} = 0$. So $\beta^*(I(\mathbf{k})) = \langle E_{j_1}, \ldots, E_{j_{n-t}} \rangle$ is a radical ideal. Hence the ideal $I(\mathbf{k})$ is also radical and generated by the $n-t$ linear functions $\gamma^*(E_{j_1}), \ldots, \gamma^*(E_{j_{n-t}})$ in the variables $U_1, \ldots, U_n$, since β^* is an isomorphism with inverse γ^*. $\diamond$

Remark 1.4.19 If $\mathbf{i} = (i_1, \ldots, i_t)$, $\mathbf{j} = (j_1, \ldots, i_u)$ and $\mathbf{k} = (k_1, \ldots, k_v)$ consist of increasing entries such that $\{i_1, \ldots, i_t\} \cap \{j_1, \ldots, i_u\} = \{k_1, \ldots, k_v\}$, then it is quite easily seen that $L(\mathbf{i}) \cap L(\mathbf{j}) = L(\mathbf{k})$.

Definition 1.4.20 Let $\mathcal{V}$ be an $l \times m$ matrix with entries in $\mathbb{F}_q[U_1, \ldots, U_n]$. Let $I(t, \mathcal{V})$ be the ideal generated by the determinants of all $(t+1) \times (t+1)$ submatrices of $\mathcal{V}_{l,t+1}$. Let $V(t, \mathcal{V})$ be the variety of $I(t, \mathcal{V})$ in $\overline{\mathbb{F}_q}^n$.

Remark 1.4.21 The ideal $I(t, \mathcal{V})$ is invariant under elementary row operations and elementary column operations on the first $t+1$ columns. In particular $I(t, \mathcal{V}) = I(t, \mathcal{SV})$ if $\mathcal{V}$ is an $l \times m$ matrix and $\mathcal{S}$ is an invertible $l \times l$ matrix. See [36].

Remark 1.4.22 Let $\mathbf{u} \in \overline{\mathbb{F}_q}^n$. The rank of $\mathcal{V}_{l,t+1}$ at $\mathbf{u}$ is at most t if and only if $\mathbf{u} \in V(t, \mathcal{V})$. See [36].

Remark 1.4.23 Let $\mathbf{u} \in \overline{\mathbb{F}_q}^n$ and $U_i = u_i$ for all i. Then $\mathbf{u} = \mathbf{u}(\mathbf{e})$ for some $\mathbf{e} \in \overline{\mathbb{F}_q}^n$ by Remark 1.4.2. Hence $U_{ij} = \mathbf{u}_{ij}(\mathbf{e})$ by Remark 1.4.16. Now $\mathbf{u} \in V(t,\mathcal{U})$ if and only if $\mathrm{rank}(\mathcal{U}_{n,t+1}(\mathbf{e})) \leq t$ by Remark 1.4.22. But $\mathrm{rank}(\mathcal{U}_{n,t+1}(\mathbf{e}))$ is equal to $\min\{t+1, \mathrm{wt}(\mathbf{e})\}$ by Proposition 1.4.13. Hence $\mathbf{u} \in V(t,\mathcal{U})$ if and only if $\mathrm{wt}(\mathbf{e}) \leq t$.

Theorem 1.4.24 *The variety $V(t,\mathcal{U})$ is the union of the $\binom{n}{t}$ irreducible components $L(\mathbf{k})$ with $1 \leq k_1 < \cdots < k_t \leq n$. Furthermore $I(t,\mathcal{U})$ is a radical ideal with*

$$I(t,\mathcal{U}) = \bigcap_{1 \leq k_1 < \cdots < k_t \leq n} I(\mathbf{k}).$$

Proof. We have that $\mathbf{u} \in V(t,\mathcal{U})$ if and only if $\mathrm{wt}(\mathbf{e}) \leq t$ by Remark 1.4.23. Now $\mathbf{u} = \mathbf{u}(\mathbf{e}) = B\mathbf{e}^T = e_1\mathbf{b}'_1 + \cdots + e_n\mathbf{b}'_n$. Furthermore $\mathrm{wt}(\mathbf{e}) \leq t$ if and only if the support of $\mathbf{e}$ is contained in $\{k_1, \ldots, k_t\}$ for some $\mathbf{k}$ with $1 \leq k_1 < \cdots < k_t \leq n$. Therefore $\mathbf{u} \in V(t,\mathcal{U})$ if and only if $\mathbf{u} \in L(\mathbf{k})$ for some t-tuple $\mathbf{k}$ with increasing entries by Lemma 1.4.18. Note that a linear space is irreducible. Hence $V(t,\mathcal{U})$ is the union of the $\binom{n}{t}$ irreducible components $L(\mathbf{k})$.

Let $D(E)$ be the diagonal matrix with the variables $E_1, \ldots, E_n$ on the diagonal. Then the componentwise application of β^* to the entries of $\mathcal{U}$ yields

$$\beta^*(\mathcal{U}) = BD(E)B^T,$$

because

$$\beta^*(U_{ij}) = \beta^*\left(\sum_{l=1}^n \mu_l^{ij} U_l\right) = \sum_{l=1}^n \sum_{l'=1}^n \mu_l^{ij} b_{ll'} E_{l'} = \sum_{l'=1}^n b_{il'}b_{jl'}E_{l'},$$

since $\mathbf{b}_i * \mathbf{b}_j = \sum_{l=1}^n \mu_l^{ij}\mathbf{b}_l$.
Remark 1.4.21 yields

$$I(t,\mathcal{U}) = I(t,\mathcal{U}_{n,t+1}) = I(t, B^{-1}\mathcal{U}_{n,t+1}).$$

Now $\beta^*(\mathcal{U}_{n,t+1}) = BD(E)B_{t+1}^T$. So $\beta^*(B^{-1}\mathcal{U}_{n,t+1}) = D(E)B_{t+1}^T$. Let $\mathbf{i}$ be an $(t+1)$-tuple with $1 \leq i_1 < \cdots < i_{t+1} \leq n$. Then the $(t+1) \times (t+1)$ minor of $\beta^*(B^{-1}\mathcal{U}_{n,t+1})$ consisting of the rows indexed by $\mathbf{i}$ is up to a nonzero scalar equal to $E_{i_1} \cdot \ldots \cdot E_{i_{t+1}}$, since all the $(t+1) \times (t+1)$ submatrices of B_{t+1} have rank $t+1$. Therefore $\beta^*(I(t,\mathcal{U}))$ is generated by all $(t+1)$-fold products $E_{i_1} \cdot \ldots \cdot E_{i_{t+1}}$. So

$$\beta^*(I(t,\mathcal{U})) = \bigcap_{1 \leq j_1 < \cdots < j_{n-t} \leq n} \langle E_{j_1}, \ldots, E_{j_{n-t}} \rangle$$

1.4. METHODS BASED ON QUADRATIC EQUATIONS 41

by Lemma 1.3.18. The inverse of the map β^* is γ^* by Remark 1.4.2. Hence

$$I(t,\mathcal{U}) = \bigcap_{1\leq j_1<\cdots<j_{n-t}\leq n} \langle \gamma^*(E_{j_1}),\ldots,\gamma^*(E_{j_{n-t}})\rangle = \bigcap_{1\leq k_1<\cdots<k_t\leq n} I(\mathbf{k})$$

by Lemma 1.4.18. Therefore $I(t,\mathcal{U})$ is radical, since it is an intersection of radical ideals. ◇

Remark 1.4.25 The ideal $I(t,\mathcal{U})$ is equal to $\gamma^*(I(t,n))$ and is generated by $\binom{n}{t+1}$ homogeneous polynomials of degree $t+1$.

Remark 1.4.26 For definitions and properties of determinantal ideals, rings and varieties we refer to the early work [112] and the more recent books [36, 60]. Determinantal rings are Cohen-Macaulay by Eagon and Hochster, see [60, Theorem 18.18]. Consequently determinantal ideals are unmixed [60, Corollary 18.14], that is all associated primes are minimal and have the same codimension, in particular there are no embedded primes [60, §3.1].
The case when the matrix $\mathcal{U}$ is "Hankel", "Toeplitz" or "Catalecticant", that is where the entries of $\mathcal{U}$ are constant along diagonals: $U_{ij} = U_{i+j-1}$, is treated in [59].

Proposition 1.4.27 *If $t < n$, then the singular locus of $V(t,\mathcal{U})$ is $V(t-1,\mathcal{U})$.*

Proof. Every component $L(\mathbf{i})$ of $V(t,\mathcal{U})$ is non-singular, since it is a linear subspace. Hence the singular locus of $V(t,\mathcal{U})$ is the union of all the intersections of two distinct components. If $\mathbf{i} = (i_1,\ldots,i_t)$ and $\mathbf{j} = (j_1,\ldots,j_t)$ consist of increasing entries, such that $\{i_1,\ldots,i_t\} \cap \{j_1,\ldots,j_t\} = \{k_1,\ldots,k_v\}$ with $\mathbf{k} = (k_1,\ldots,k_v)$, then $L(\mathbf{i}) \cap L(\mathbf{j}) = L(\mathbf{k})$ by Remark 1.4.19. If moreover $\mathbf{i} \neq \mathbf{j}$, then $v \leq t-1$ and $L(\mathbf{i}) \cap L(\mathbf{j}) \subseteq V(t-1,\mathcal{U})$.
Conversely, let $L(\mathbf{k})$ be a component of $V(t-1,\mathcal{U})$ with $\mathbf{k} = (k_1,\ldots,k_{t-1})$. Then $\{i_1,\ldots,i_t\} \cap \{j_1,\ldots,j_t\} = \{k_1,\ldots,k_{t-1}\}$ for some $\mathbf{i} = (i_1,\ldots,i_t)$ and $\mathbf{j} = (j_1,\ldots,j_t)$, since $t < n$. Hence $L(\mathbf{k}) = L(\mathbf{i}) \cap L(\mathbf{j})$ is in the singular locus of $V(t,\mathcal{U})$. ◇

Example 1.4.28 Let B be the RS matrix of Example 1.4.14. So $\mathbf{b}'_1 = (1,1,1)^T$, $\mathbf{b}'_2 = (1,\alpha,\alpha^2)^T$ and $\mathbf{b}'_3 = (1,\alpha^2,\alpha)^T$ are the columns of B. The matrix $\mathcal{U}$ is of the form
$$\begin{pmatrix} U_1 & U_2 & U_3 \\ U_2 & U_3 & U_1 \\ U_3 & U_1 & U_2 \end{pmatrix}.$$

If $t = 0$, then $V(0, \mathcal{U})$ consists of the origin and indeed

$$I(0, \mathcal{U}) = \langle U_1, U_2, U_3 \rangle.$$

For $t = 1$ we have that $V(1, \mathcal{U})$ is the union of the lines $L(1)$, $L(2)$ and $L(3)$ through the origin with directions $\mathbf{b}'_1$, $\mathbf{b}'_2$ and $\mathbf{b}'_3$, respectively and

$$I(1, \mathcal{U}) = \langle U_2^2 - U_1 U_3, U_1^2 - U_2 U_3, U_3^2 - U_1 U_2 \rangle.$$

In case $t = 2$ we have that $V(2, \mathcal{U})$ is the union of the planes $L(1, 2)$, $L(1, 3)$ and $L(2, 3)$, where $L(i, j)$ is the plane through the origin generated by $\mathbf{b}'_i$ and $\mathbf{b}'_j$. The corresponding ideals are $I(1, 2) = \langle U_1 + \alpha U_2 + \alpha^2 U_3 \rangle$, $I(1, 3) = \langle U_1 + \alpha^2 U_2 + \alpha U_3 \rangle$ and $I(2, 3) = \langle U_1 + U_2 + U_3 \rangle$, respectively. Furthermore

$$I(2, \mathcal{U}) = \langle U_1^3 + U_2^3 + U_3^3 - 3 U_1 U_2 U_3 \rangle.$$

Indeed we have that

$$I(2, \mathcal{U}) = I(1, 2) \cap I(1, 3) \cap I(2, 3),$$

since

$$U_1^3 + U_2^3 + U_3^3 - 3 U_1 U_2 U_3 = (U_1 + \alpha U_2 + \alpha^2 U_3)(U_1 + \alpha^2 U_2 + \alpha U_3)(U_1 + U_2 + U_3).$$

Definition 1.4.29 The ideal $I(t, \mathcal{U}, V)$ in the ring $\mathbb{F}_q[U_1, \ldots, U_n, V_1, \ldots, V_t]$ is generated by the elements

$$\sum_{j=1}^{t} U_{ij} V_j \ - \ U_{i, t+1} \quad \text{for} \quad i = 1, \ldots, n$$

Let $V(t, \mathcal{U}, V)$ be the zero set of $I(t, \mathcal{U}, V)$ over $\overline{\mathbb{F}_q}$.

Remark 1.4.30 For every $\mathbf{u}$ there is a unique $\mathbf{e}$ such that $\mathbf{u} = \mathbf{u}(\mathbf{e})$ by Remark 1.4.2. Let $\mathcal{U} = \mathcal{U}(\mathbf{e})$. Then $(\mathbf{u}, \mathbf{v})$ is an element of $V(t, \mathcal{U}, V)$ for some $\mathbf{v}$ if and only if column $t + 1$ of $\mathcal{U}$ is a linear combination of the first t columns of $\mathcal{U}$.

Lemma 1.4.31 Let $\mathbf{u} = \mathbf{u}(\mathbf{e})$ (and $\mathcal{U} = \mathcal{U}(\mathbf{e})$). If $\mathbf{u} \in V(t, \mathcal{U})$, then there is a $t' \leq t$ and a $\mathbf{v}$ such that $(\mathbf{u}, \mathbf{v}) \in V(t', \mathcal{U}, V)$.

Proof. Suppose $\mathbf{u} \in V(t, \mathcal{U})$. Then $\text{rank}(\mathcal{U}_{n, t+1}(\mathbf{e})) \leq t$ by Remark 1.4.23. So the first $t + 1$ columns of $\mathcal{U}$ are dependent. Hence there is a $t' \leq t$ such that column $t' + 1$ of $\mathcal{U}$ is a linear combination of the first t' columns of $\mathcal{U}$. Therefore $(\mathbf{u}, \mathbf{v})$ is an element $V(t', \mathcal{U}, V)$ for some $\mathbf{v}$, by Remark 1.4.30. ◇

1.4. METHODS BASED ON QUADRATIC EQUATIONS

Proposition 1.4.32
$$I(t,\mathcal{U}) \subseteq I(t,\mathcal{U},V).$$

Proof. Let R_t be the factor ring $\mathbb{F}_q[U_1,\ldots,U_n,V_1,\ldots,V_t]/I(t,\mathcal{U},V)$. Then the following equations hold in the ring R_t

$$\sum_{l=1}^{t} U_{il}V_l - U_{i,t+1} = 0$$

for all $i = 1,\ldots,n$. That is

$$\mathcal{U}_{n,t}\mathbf{V} = \mathbf{U_{t+1}},$$

where $\mathbf{V}$ is the column vector $(V_1,\ldots,V_t)^T$ and $\mathbf{U_{t+1}}$ is the column vector $(U_{1,t+1},\ldots,U_{n,t+1})^T$. Let $\mathbf{i} = (i_1,\ldots,i_t,i_{t+1})$ with $1 \leq i_1 < \cdots < i_t < i_{t+1} \leq n$. Let N be the $(t+1) \times (t+1)$ matrix with entries $n_{jl} = U_{i_j,l}$ for $1 \leq j,l \leq t+1$. Now $I(t,\mathcal{U})$ is generated by elements of the form $\det(N)$ for all possible choices of $\mathbf{i}$, and we claim that $\det(N) \in I(t,\mathcal{U},V)$.

For the following properties of determinants of matrices over a commutative ring with unit element we refer to [86, XIII, §4]. Let N_{jl} be the $t \times t$ minor of N obtained by deleting the j-th row and the l-th column of N. Let $M = N_{t+1,t+1}$. Let M_{jl} be the $(t-1) \times (t-1)$ minor of M obtained by deleting the j-th row and the l-th column of M. Let $\mathrm{ad}(M)$ be the $t \times t$ adjoint matrix of M with the entry $(-1)^{j+l}\det(M_{lj})$ at the j-th row and the l-th column. Then

$$\mathrm{ad}(M)M = M \cdot \mathrm{ad}(M) = \det(M)I_t.$$

Let $\mathbf{b}$ be the column vector $(U_{i_1,t+1},\ldots,U_{i_t,t+1})^T$. Then $M\mathbf{V} = \mathbf{b}$. So

$$\det(M)\mathbf{V} = \mathrm{ad}(M)M\mathbf{V} = \mathrm{ad}(M)\mathbf{b}.$$

Hence

$$\det(M)V_l = \sum_{j=1}^{t}(-1)^{j+l}\det(M_{jl})U_{i_j,t+1}$$

for all $l = 1,\ldots,t$. Now the right hand side is a cofactor expansion of the determinant of $N_{t+1,l}$ along the last column

$$\det(N_{t+1,l}) = \sum_{j=1}^{t}(-1)^{t+j}\det(M_{jl})U_{i_j,t+1}.$$

Furthermore $M = N_{t+1,t+1}$. So

$$\det(N_{t+1,t+1})V_l = (-1)^{t+l}\det(N_{t+1,l})$$

for all $l = 1, \ldots, t$. Multiplying the equation at the beginning for $i = i_{t+1}$ by $-\det(N_{t+1,t+1})$ and using the last equations yields

$$\sum_{l=1}^{t}(-1)^{t+1+l}\det(N_{t+1,l})U_{i_{t+1}l} + \det(N_{t+1,t+1})U_{i_{t+1},t+1} = 0.$$

But the left hand side of the equation is the cofactor expansion of $\det(N)$ along row $t+1$. So $\det(N) = 0$ in R_t. Hence $\det(N) \in I(t,\mathcal{U},V)$ for every i. This proves the claim and therefore $I(t,\mathcal{U}) \subseteq I(t,\mathcal{U},V)$. ⋄

Theorem 1.4.33 *The ideal $I(t,\mathcal{U},V)$ is radical and the algebraic set $V(t,\mathcal{U},V)$ is a complete intersection and consists of $\sum_{w=0}^{t}\binom{n}{w}$ components all of dimension t.*

Proof. Let the map $\varphi : \overline{\mathbb{F}}^n \times \overline{\mathbb{F}}^t \to \overline{\mathbb{F}}^n$ be the projection on the first factor. Now $I(t,\mathcal{U})$ is contained in $I(t,\mathcal{U},V)$ by Proposition 1.4.32. Hence the projection map restricts to a map $\varphi : V(t,\mathcal{U},V) \to V(t,\mathcal{U})$. Consider the following diagram.

$$\begin{array}{ccc}
& V(t,\mathcal{U},V) & \subset \overline{\mathbb{F}}^n \times \overline{\mathbb{F}}^t \\
& \downarrow \varphi & \downarrow \\
V(t-1,\mathcal{U}) \subset & V(t,\mathcal{U}) & \subset \overline{\mathbb{F}}^n
\end{array}$$

Let $\mathbf{u} \in V(t,\mathcal{U})$. Then there is a w such that $\mathbf{u} \in V(w,\mathcal{U}) \setminus V(w-1,\mathcal{U})$. Then $\mathbf{u} = u(\mathbf{e})$ and $\mathrm{wt}(\mathbf{e}) = w$ and $\mathrm{rank}(\mathcal{U}_{n,w}(\mathbf{e})) = w$ by Remark 1.4.23. So the first w columns of $\mathcal{U}_{n,t+1}(\mathbf{e})$ are independent. Gaussian elimination of $\mathcal{U}_{n,t+1}(\mathbf{e})$ gives that the row reduced echelon form of this matrix is of the form

$$\left(\begin{array}{c|c} I_w & B \\ \hline 0 & 0 \end{array}\right).$$

Hence the variables $V_{w+1},\ldots,V_t$ are free to choose and $V_1,\ldots,V_w$ are uniquely determined by these free variables. Note that $\mathcal{U}_{n,t+1}$ is an $n \times (t+1)$ matrix and is an extended matrix of the linear system (w.r.t V_j's) written in Definition 1.4.29; so that only variables $V_1,\ldots,V_t$ are considered. Therefore $\varphi^{-1}(\mathbf{u})$ has dimension $t - w$.

Furthermore $\mathbf{u}$ is in the component $L(\mathbf{k})$ of $Z(w,\mathcal{U})$ of dimension w by Theorem 1.4.24 and $\mathbf{u}$ is exactly in this component by Proposition 1.4.27. Now there exists a $w \times w$ submatrix of $\mathcal{U}_{n,w}$ with determinant Δ which is a polynomial in $U_1,\ldots,U_n$ such that $\Delta(\mathbf{u})$ is not zero, since $\mathrm{rank}(\mathcal{U}_{n,w}(\mathbf{e})) = w$. Let S_w be the factor ring $\mathbb{F}[U_1,\ldots,U_n,V_1,\ldots,V_t]/I(w,\mathcal{U})$. Let $S_w[\Delta^{-1}]$ be the localization of the ring S_w at the multiplicative set consisting of powers

1.4. METHODS BASED ON QUADRATIC EQUATIONS 45

of Δ, see [60, §2]. Then Gaussian elimination of $\mathcal{U}_{n,t+1}$ over the ring $S_w[\Delta^{-1}]$ gives again that the row reduced echelon form implies that the variables $V_{w+1}, \ldots, V_t$ are free to choose and $V_1, \ldots, V_w$ are uniquely determined by these free variables. This is readily seen as in this case only division by Δ is needed, cf. [86], chapter XIII, Proposition 4.16.

So $L(\mathbf{k}) \setminus V(\Delta)$ is an open dense neighborhood of $\mathbf{u}$ in $L(\mathbf{k})$ in the Zariski topology such that $\varphi^{-1}(L(\mathbf{k}) \setminus V(\Delta))$ is irreducible and has dimension t. Hence every component of $V(t, \mathcal{U}, V)$ has dimension t. Moreover there are $\binom{n}{w}$ such components above $V(w, \mathcal{U})$ that are not above $V(w-1, \mathcal{U})$ by Theorem 1.4.24. Therefore $V(t, \mathcal{U}, V)$ consists of $\sum_{w=0}^{t} \binom{n}{w}$ components. Finally all components of $V(t, \mathcal{U}, V)$ have dimension t (equivalently, codimension n) and $V(t, \mathcal{U}, V)$ is the zero set in affine space of dimension $n + t$ of the ideal $I(t, \mathcal{U}, V)$ that is generated by n elements. So $V(t, \mathcal{U}, V)$ is an ideal theoretic complete intersection by [85, Definition 3.12 of Chap. V]. This implies that $I(t, \mathcal{U}, V)$ is generated by some regular sequence, so $V(t, \mathcal{U}, V)$ is a complete intersection, and all its associated primes have codimension n, by [85, Proposition 3.14 of Chap. VI].

Let P be an associated prime of $I(t, \mathcal{U}, V)$. Then $V(P)$ is a component of $V(t, \mathcal{U}, V)$. It was shown above that there exists a determinant Δ of a $w \times w$ submatrix of $\mathcal{U}_{n,w}$ such that $\Delta \notin P$. Let $R = \mathbb{F}[U_1, \ldots, U_n, V_1, \ldots, V_t]$ and R_P the localization of R at P. Then $V_{w+1}, \ldots, V_t$ are free to choose and $V_1, \ldots, V_w$ are uniquely determined by these free variables in R_P. Let $S = \mathbb{F}[U_1, \ldots, U_n]$ and $Q = \Gamma \cap S$. Then Q is an associated prime of $I(w, \mathcal{U})$. Let S_Q be the localization of S at Q. Then $R_P/I(t, \mathcal{U}, V)R_P \cong (S_Q/QS_Q)[V_{w+1}, \ldots, V_t]$ is an integral domain. Hence $I(t, \mathcal{U}, V)R_P$ is a prime ideal. Therefore $I(t, \mathcal{U}, V)R_P = PR_P$ is prime for all associated primes of $I(t, \mathcal{U}, V)$. So all primary ideals of $I(t, \mathcal{U}, V)$ are prime. Therefore, the ideal $I(t, \mathcal{U}, V)$ is radical. ◇

Example 1.4.34 This is a continuation of Example 1.4.28. Suppose that $(\mathbf{u}, \mathbf{v})$ is a solution of the following system of equations.

$$\begin{cases} U_1V_1 + U_2V_2 = U_3, \\ U_2V_1 + U_3V_2 = U_1, \\ U_3V_1 + U_1V_2 = U_2. \end{cases}$$

0) If $\mathbf{u} \in V(0, \mathcal{U})$, then $\mathbf{u} = 0$. In this case $\mathbf{v}$ is free to choose and the component above $V(0, \mathcal{U})$ is the set $\{0\} \times \overline{\mathbb{F}}_q^2$.

1) If $\mathbf{u} \in V(1, \mathcal{U}) \setminus V(0, \mathcal{U})$, then $u_i \neq 0$ and $u_i^2 = u_{i-1}u_{i+1}$ for all i where the indices are counted modulo 3. Gaussian elimination of the extended matrix

associated with the system of equations gives

$$\begin{pmatrix} u_1 & u_2 & u_3 \\ u_2 & u_3 & u_1 \\ u_3 & u_1 & u_2 \end{pmatrix} \sim \begin{pmatrix} 1 & u_2/u_1 & u_3/u_1 \\ 0 & (u_1u_3 - u_2^2)/u_1 & (u_1^2 - u_2u_3)/u_1 \\ 0 & (u_1^2 - u_2u_3)/u_1 & (u_1u_2 - u_3^2)/u_1 \end{pmatrix}.$$

The last two rows are in fact zero. Hence $v_1 = (u_3 - u_2v_2)/u_1$, where v_2 is free to choose.

2) Now suppose that $\mathbf{u} \in V(2, \mathcal{U}) \setminus V(1, \mathcal{U})$. Let $d_i = u_i^2 - u_{i-1}u_{i+1}$ and $d = u_1d_1 + u_2d_2 + u_3d_3$. Then $d = 0$ and $d_i \neq 0$ for some i. Suppose for instance that $d_2 \neq 0$. Then Gaussian elimination gives.

$$\begin{pmatrix} u_1 & u_2 & u_3 \\ u_2 & u_3 & u_1 \\ u_3 & u_1 & u_2 \end{pmatrix} \sim \begin{pmatrix} 1 & 0 & -d_3/d_2 \\ 0 & 1 & -d_1/d_2 \\ 0 & 0 & d/d_2 \end{pmatrix}.$$

The element in the right lower corner is $d/d_2 = 0$. Hence the unique solution for $\mathbf{v}$ is given by $v_1 = -d_3/d_2$ and $v_2 = -d_1/d_2$.

Proposition 1.4.35

$$I(t, \mathcal{U}, V) \cap \mathbb{F}[U_1, \ldots, U_n] = I(t, \mathcal{U}).$$

Proof. Indeed, from Proposition 1.4.32 we have that $I(t, \mathcal{U}) \subseteq I(t, \mathcal{U}, V)$ and thus $I(t, \mathcal{U}) \subseteq I(t, \mathcal{U}, V) \cap \mathbb{F}[U_1, \ldots, U_n]$. The map $\varphi : \overline{\mathbb{F}}^n \times \overline{\mathbb{F}}^t \to \overline{\mathbb{F}}^n$ is the projection on the first factor. From the proof of Theorem 1.4.33 it follows that $\varphi(V(t, \mathcal{U}, V)) = V(t, \mathcal{U})$. Now $I(t, \mathcal{U}, V) \cap \mathbb{F}[U_1, \ldots, U_n]$ is the elimination ideal of $I(t, \mathcal{U}, V)$ with respect to the projection φ. So $V(I(t, \mathcal{U}, V) \cap \mathbb{F}[U_1, \ldots, U_n])$ is the Zariski-closure of $\varphi(V(t, \mathcal{U}, V))$. See [55, Chap. 3, §2], [60, Chap. 14] and [60, Proposition 15.30]. Hence $V(I(t, \mathcal{U}, V) \cap \mathbb{F}[U_1, \ldots, U_n]) = V(t, \mathcal{U})$. Therefore

$$\mathrm{rad}(I(t, \mathcal{U}, V) \cap \mathbb{F}[U_1, \ldots, U_n]) = \mathrm{rad}(I(t, \mathcal{U})).$$

But $I(t, \mathcal{U})$ is radical due to Theorem 1.4.24. Hence $I(t, \mathcal{U}, V) \cap \mathbb{F}[U_1, \ldots, U]$ is also radical and equal to $I(t, \mathcal{U})$. ◇

1.4.3 Nearest codeword decoding

Without loss of generality we may assume, after a finite extension of the finite field $\mathbb{F}_q$, that $n \leq q$. Let $\mathbf{b}_1, \ldots, \mathbf{b}_n$ be a basis of $\mathbb{F}_q^n$. From now on we assume that the corresponding matrix B is an MDS matrix.

1.4. METHODS BASED ON QUADRATIC EQUATIONS

Let C be an $\mathbb{F}_q$-linear code with parameters $[n, k, d]$. Choose a parity check matrix H of C. The redundancy is $r = n - k$. Let $\mathbf{h}_1, \ldots, \mathbf{h}_r$ be the rows of H. The row $\mathbf{h}_i$ is a linear combination of the basis $\mathbf{b}_1, \ldots, \mathbf{b}_n$, that is there are constants $a_{ij} \in \mathbb{F}_q$ such that

$$\mathbf{h}_i = \sum_{j=1}^{n} a_{ij} \mathbf{b}_j.$$

In other words $H = AB$ where A is the $r \times n$ matrix with entries a_{ij}.

Remark 1.4.36 Let $\mathbf{r} = \mathbf{c} + \mathbf{e}$ be a received word with $\mathbf{c} \in C$ a codeword and $\mathbf{e}$ an error vector. The syndromes of $\mathbf{r}$ and $\mathbf{e}$ with respect to H are equal and known: $s_i(\mathbf{r}) := \mathbf{h}_i \cdot \mathbf{r} = \mathbf{h}_i \cdot \mathbf{e} = s_i(\mathbf{e})$ as noted in Remark 1.3.12 and they can be expressed in the unknown syndromes of $\mathbf{e}$ with respect to B:

$$s_i(\mathbf{r}) = \sum_{j=1}^{n} a_{ij} u_j(\mathbf{e}),$$

since $\mathbf{h}_i = \sum_{j=1}^{n} a_{ij} \mathbf{b}_j$ and $\mathbf{b}_j \cdot \mathbf{e} = u_j(\mathbf{e})$.

Definition 1.4.37 The ideal $J(\mathbf{r})$ in the ring $\mathbb{F}_q[U_1, \ldots, U_n]$ is generated by the elements

$$\sum_{l=1}^{n} a_{jl} U_l - s_j(\mathbf{r}) \quad \text{for} \quad j = 1, \ldots, r.$$

Let $J(t, \mathbf{r})$ be the ideal in $\mathbb{F}_q[U_1, \ldots, U_n, V_1, \ldots, V_t]$ generated by $J(\mathbf{r})$ and $I(t, \mathcal{U}, V)$.

Remark 1.4.38 The ideal $J(t, \mathbf{r})$ is generated by $n - k$ linear functions and n quadratic polynomials. In principle, we can also express some $n - k$ variables via k others using the parity check matrix H, and then substitute in the quadratic part. In this way we obtain an ideal generated by n quadratic polynomials in $k + t$ variables.

Lemma 1.4.39 If $\mathbf{r} = \mathbf{c} + \mathbf{e}$ for some $\mathbf{c} \in C$ and $wt(\mathbf{e}) = t$, then there is a $\mathbf{v}$ such that $(\mathbf{u}(\mathbf{e}), \mathbf{v})$ is a solution of $J(t, \mathbf{r})$.

Proof. Let $\mathbf{r} = \mathbf{c} + \mathbf{e}$ for some $\mathbf{c} \in C$ and $\mathbf{u} = \mathbf{u}(\mathbf{e})$. Then $\mathbf{u}$ is a solution of $J(\mathbf{r})$ by Remark 1.4.36. Now $\mathcal{U} = \mathcal{U}(\mathbf{e})$ and $\text{rank}(\mathcal{U}_{nv}(\mathbf{e})) = \min\{v, wt(\mathbf{e})\}$ by Proposition 1.4.13. Let $wt(\mathbf{e}) = t$. Then $\text{rank}(\mathcal{U}_{nv}) = v$ if $v < t$ and $\text{rank}(\mathcal{U}_{nv}) = t$ if $v \geq t$. So column $t+1$ of $\mathcal{U}_{n,t+1}$ is a linear combination of the first t columns of $\mathcal{U}$. Hence there is a $\mathbf{v}$ such that $(\mathbf{u}, \mathbf{v})$ is a solution of $I(t, \mathcal{U}, V)$, by Remark 1.4.30. Hence $(\mathbf{u}, \mathbf{v})$ is a solution of $J(t, \mathbf{r})$. ◇

Lemma 1.4.40 *Let* $(\mathbf{u}, \mathbf{v})$ *be a solution of* $J(t, \mathbf{r})$. *Then there is a unique* $\mathbf{e}$ *of weight at most* t *such that* $\mathbf{u} = \mathbf{u}(\mathbf{e})$, *furthermore* $\mathbf{r} = \mathbf{c} + \mathbf{e}$ *for some* $\mathbf{c}$ *in* $\mathbb{F}_{q^m} C$ *for some* m.

Proof. Let $(\mathbf{u}, \mathbf{v})$ be a solution of $J(t, \mathbf{r})$. Then there is a unique $\mathbf{e}$ such that $\mathbf{u} = \mathbf{u}(\mathbf{e})$ by Remark 1.4.2.
The element $\mathbf{u}$ in $\mathbb{F}_{q^m}^n$ for some m is a solution of $J(\mathbf{r})$. Hence $s_i(\mathbf{r}) = \sum_{j=1}^n a_{ij} u_j(\mathbf{e})$ for all i. But $s_i(\mathbf{e}) = \sum_{j=1}^n a_{ij} u_j(\mathbf{e})$ for all i, and $\mathbf{u} = \mathbf{u}(\mathbf{e})$. So $\mathbf{s}(\mathbf{r} - \mathbf{e}) = 0$. Hence there is a $\mathbf{c}$ in $\mathbb{F}_{q^m} C$ such that $\mathbf{r} = \mathbf{c} + \mathbf{e}$.
Now $(\mathbf{u}, \mathbf{v})$ is a solution of $I(t, \mathcal{U}, V)$. Hence the $(t+1)$-th column of $\mathcal{U}(\mathbf{e})$ is a linear combination of the first t columns of $\mathcal{U}(\mathbf{e})$, by Remark 1.4.30. Therefore $\text{rank}(\mathcal{U}_{n,t+1}(\mathbf{e})) \leq t$. But $\text{rank}(\mathcal{U}_{n,t+1}(\mathbf{e})) = \min\{t+1, \text{wt}(\mathbf{e})\}$ by Proposition 1.4.13. Hence $\text{wt}(\mathbf{e}) \leq t$. ◇

Lemma 1.4.41 *If* $(\mathbf{u}, \mathbf{v})$ *and* $(\mathbf{u}, \mathbf{w})$ *are distinct solutions of* $J(t, \mathbf{r})$, *then there is a solution* $(\mathbf{u}, \mathbf{z})$ *of* $J(t', \mathbf{r})$ *for some* t' *with* $t' < t$. *If furthermore* $t' = 0$, *then* $\mathbf{r}$ *in* $\mathbb{F}_{q^m} C$ *for some* m.

Proof. Let $(\mathbf{u}, \mathbf{v})$ and $(\mathbf{u}, \mathbf{w})$ be distinct solutions of $J(t, \mathbf{r})$. Then $\mathbf{u}$ is a solution of $J(\mathbf{r})$ and $(\mathbf{u}, \mathbf{v})$ and $(\mathbf{u}, \mathbf{w})$ are solutions of $I(t, \mathcal{U}, V)$. There is a unique $\mathbf{e}$ such that $\mathbf{u} = \mathbf{u}(\mathbf{e})$. Hence

$$\sum_{j=1}^t u_{ij} v_j = u_{i,t+1} \quad \text{and} \quad \sum_{j=1}^t u_{ij} w_j = u_{i,t+1} \text{ for all } i.$$

So

$$\sum_{j=1}^t u_{ij}(v_j - w_j) = 0 \text{ for all } i,$$

and $\mathbf{v} - \mathbf{w} \neq 0$, since $\mathbf{v} \neq \mathbf{w}$. Hence the first t columns of the matrix $\mathcal{U}(\mathbf{e})$ are linearly dependent. Therefore there is a $t' < t$ such that column $t' + 1$ is a linear combination of the first t' columns. In other words there is a solution $(\mathbf{u}, \mathbf{z})$ of $J(t', \mathbf{r})$ by Remark 1.4.30.
If $t' = 0$, then $\mathbf{u} = \mathbf{u}(\mathbf{e})$ for a unique $\mathbf{e}$ and $\mathbf{r} = \mathbf{c} + \mathbf{e}$ for some $\mathbf{c}$ in $\mathbb{F}_{q^m} C$ and $\text{wt}(\mathbf{e}) \leq 0$, by Lemma 1.4.40. Hence $\mathbf{r}$ in $\mathbb{F}_{q^m} C$ for some m. ◇

Theorem 1.4.42 *Let* B *be an MDS matrix with structure constants* μ_l^{ij} *and linear functions* U_{ij}. *Let* H *be a parity check matrix of the code* C *such that* $H = AB$. *Let* $\mathbf{r} \in \mathbb{F}_q^n$ *and* $\mathbf{r} \notin C$. *Let* t *be the smallest non-negative integer such that* $J(t, \mathbf{r})$ *has a solution* $(\mathbf{u}, \mathbf{v})$ *over* $\overline{\mathbb{F}_q}$. *Then* $t = d(\mathbf{r}, C)$. *Furthermore for every* $\mathbf{c} \in \mathcal{L}(\mathbf{r}, C)$ *there is a unique solution* $(\mathbf{u}, \mathbf{v})$ *of* $J(t, \mathbf{r})$ *over* $\overline{\mathbb{F}_q}$ *such that* $\mathbf{u} = \mathbf{u}(\mathbf{r} - \mathbf{c})$.

1.4. METHODS BASED ON QUADRATIC EQUATIONS

Proof. Take $\mathbf{c} \in \mathcal{L}(\mathbf{r}, C)$, then $\mathbf{r} = \mathbf{c} + \mathbf{e}, \mathbf{c} \in C, \mathbf{e} \in \mathbb{F}_q^n$ with $\mathrm{wt}(\mathbf{e}) = d(\mathbf{r}, C) := w$. So by Lemma 1.4.39 there is a solution $(\mathbf{u}, \mathbf{v})$ of $J(w, \mathbf{r})$, such that $\mathbf{u} = \mathbf{u}(\mathbf{e}) = \mathbf{u}(\mathbf{r} - \mathbf{c})$. Since t is the smallest, such that $J(t, \mathbf{r})$ has a solution it follows that $t \leq w$.

Let $(\tilde{\mathbf{u}}, \tilde{\mathbf{v}})$ be a solution of $J(t, \mathbf{r})$. By Lemma 1.4.40 we have that $\mathbf{r} = \tilde{\mathbf{c}} + \tilde{\mathbf{e}}$, where $\tilde{\mathbf{c}} \in \tilde{C} := \mathbb{F}_{q^m} C$ for some m and $\tilde{\mathbf{e}} \in \mathbb{F}_{q^m}^n$ with $\mathrm{wt}(\tilde{\mathbf{e}}) \leq t$. Now $\mathrm{wt}(\mathbf{e}) \leq t \leq w = d(\mathbf{r}, C) = d(\mathbf{r}, \tilde{C})$ by Proposition 1.3.14. On the other hand, since $\mathbf{r} = \tilde{\mathbf{c}} + \tilde{\mathbf{e}}$ it follows that $\mathrm{wt}(\tilde{\mathbf{e}}) \geq d(\mathbf{r}, \tilde{C})$. Combining with the former we have that $t \geq \mathrm{wt}(\tilde{\mathbf{e}}) \geq d(\mathbf{r}, \tilde{C}) = w$. From the minimality of t it follows then that $t = w$.

Now let $(\mathbf{u}', \mathbf{v}')$ be another solution of $J(t, \mathbf{r}) = J(w, \mathbf{r})$ such that $\mathbf{u}' = \mathbf{u}(\mathbf{r} - \mathbf{c}) = \mathbf{u}$, so $(\mathbf{u}', \mathbf{v}') = (\mathbf{u}, \mathbf{v}')$ with $\mathbf{v} \neq \mathbf{v}'$. By Lemma 1.4.41 there exists a solution of $J(t', \mathbf{r})$ for $t' < t$. This contradicts the minimality of t. ◇

For the further arguments we need the following two results.

Proposition 1.4.43 Let $I \subset \mathbb{F}[X_1, \ldots, X_n]$ for some field $\mathbb{F}$ be a zero-dimensional ideal. Then I is radical iff $I \cap \mathbb{F}[X_i] = \langle f_i \rangle, i = 1, \ldots, n$, where all f_i are square-free polynomials.

Proof. See [75, Proposition 4.5.1] ◇

Proposition 1.4.44 Let $I_i, i = 1, \ldots, m$ and J be radical ideals in the polynomial ring $R := \mathbb{F}[X_1, \ldots, X_n]$ for some field $\mathbb{F}$. Let $I_i + J$ be radical for all $i = 1, \ldots, m$. Suppose that $I_i + I_j + J = R$ for all mutually distinct i and j. Let $I = \bigcap_{i=1}^m I_i$. Then $I + J$ is radical.

Proof. It is enough to prove the following:

$$I + J = \bigcap_{i=1}^m (I_i + J).$$

Indeed, if we have this, then $I_i + J$ is radical for every i and $\bigcap_{i=1}^m (I_i + J)$ is radical as an intersection of radical ideals. Now to prove the above equality, we employ the Chinese Remainder Theorem, cf. [86]. Let us pass to the ring $\bar{R} := R/J$. In this ring for every two distinct i and j we have $\bar{I}_i + \bar{I}_j = \bar{R}$, where $\bar{I}_i$ and $\bar{I}_j$ are canonical images of I_i and I_j in $\bar{R}$ resp. Therewith we obtain

$$\bar{R}/\bigcap_{i=1}^m \bar{I}_i \cong \prod_{i=1}^m \bar{R}/\bar{I}_i.$$

Going back to R we have

$$R/(I + J) \cong \prod_{i=1}^m R/(I_i + J).$$

Now the element 0 on the left side comes from $I + J$ and it corresponds to the element $(0, \ldots, 0)$ on the right side that comes from $\cap_{i=1}^{m}(I_i + J)$. So we have the desired equality because of the isomorphism. ◇

Theorem 1.4.45 (Bounded distance decoding) *Let $\mathbf{r}$ be a received word, and $\mathbf{r} = \mathbf{c}+\mathbf{e}$, where $\mathbf{c} \in C$ and $wt(\mathbf{e}) = t \leq (d(C)-1)/2$. Then $J(t, \mathbf{r}) = \langle \mathcal{G} \rangle$, where $\mathcal{G} = \{U_i - u_i, V_j - v_j\}_{1 \leq i \leq n, 1 \leq j \leq t}$, where $(u_1, \ldots, u_n, v_1, \ldots, v_t)$ is a unique element of $V(t, \mathbf{r})$. Moreover $\mathcal{G}$ is the reduced Gröbner basis of $J(t, \mathbf{r})$ w.r.t any monomial order.*

Proof. Due to Theorem 1.4.42, the variety $V(t, \mathbf{r})$ has only one element $(\mathbf{u}, \mathbf{v})$, where $\mathbf{u} = (u_1, \ldots, u_n), \mathbf{v} = (v_1, \ldots, v_t)$. By Proposition 1.4.32 we have that $I(t, \mathcal{U}) + J(\mathbf{r}) \subseteq J(t, r) \cap \mathbb{F}_q[U_1, \ldots, U_n]$. Further we have $(\mathbf{u}, \mathbf{v}) \in V(t, \mathcal{U}, V)$. Hence $\mathbf{u} \in V(t, \mathcal{U})$ by Proposition 1.4.32. So there is a $\mathbf{k}$ with $1 \leq k_1 < \cdots < k_t \leq n$ such that $\mathbf{u} \in L(\mathbf{k})$ by Theorem 1.4.24. If there is another such $\mathbf{k}'$ with $\mathbf{u} \in L(\mathbf{k}')$, then $\mathbf{u} \in V(t-1, \mathcal{U})$, by Proposition 1.4.27. Hence there is a $t' < t$ and a $\mathbf{v}'$ such that $(\mathbf{u}, \mathbf{v}') \in V(t', \mathcal{U}, V)$, by Lemma 1.4.31. But this contradicts the minimality of t. So $\mathbf{u}$ is not an element of $V(t-1, \mathcal{U})$ and the $\mathbf{k}$ is unique. This means that $V(\mathbf{r})$ does not intersect $V(t, \mathcal{U})$ in singular points, which lie exactly in the mutual intersections of the linear components. Moreover, $J(\mathbf{r})$ as well as all $I(\mathbf{k})$ are generated by linear polynomials, thus $I(\mathbf{k})+J(\mathbf{r})$ is radical for all increasing t-tuples $\mathbf{k}$. So all the assumptions of Proposition 1.4.44 are satisfied and thus we have that $I(t, \mathcal{U})+J(\mathbf{r})$ is radical. It is also readily seen that $V(J(t, \mathbf{r}) \cap \mathbb{F}_q[U_1, \ldots, U_n]) = V(t, \mathcal{U}) \cap V(\mathbf{r})$, see the proof of Proposition 1.4.35. Therefore we have that $I := J(t, \mathbf{r}) \cap \mathbb{F}_q[U_1, \ldots, U_n] = I(t, \mathcal{U}) + J(\mathbf{r})$ and is radical.

Now $J(t, \mathbf{r})$ and thus I are zero-dimensional ideals. Therefore, we have $I \cap \mathbb{F}_q[U_i] = \langle f_i \rangle, i = 1, \ldots, n$, where f_i are square-free polynomials, cf. Proposition 1.4.43. Since $\mathbf{u}$ is the only element in $V(J(t, \mathbf{r}) \cap \mathbb{F}_q[U_1, \ldots, U_n])$, we have that $f_i = U_i - u_i, i = 1, \ldots, n$. So $U_i - u_i \in J(t, \mathbf{r}), i = 1, \ldots, n$. Now we will show that $J(t, \mathbf{y}) = \langle \mathcal{G} \rangle$ by computing modulo $\mathcal{M}_\mathbf{u} := \langle U_i - u_i \rangle_{i=1, \ldots, n}$. The element $\sum_{j=1}^{t} U_{ij} V_j - U_{i,t+1}$ is in $J(t, \mathbf{y})$, and $U_{ij} \equiv u_{ij}$ modulo $\mathcal{M}_\mathbf{u}$ and $\mathbf{v}$ is the unique solution of the linear equations

$$\sum_{j=1}^{t} u_{ij} V_j - u_{i,t+1} = 0, \text{ for all } i = 1, \ldots, n.$$

Hence $V_j - v_j \in J(t, \mathbf{y})$ for all $j = 1, \ldots, t$, by Gaussian elimination.
The claim on the Gröbner basis is trivial. ◇

Theorem 1.4.46 (Nearest codeword decoding) *Let $\mathbf{r}$ be a received word. Let t be the smallest positive integer such that $J(t, \mathbf{y})$ has a solution. Let*

1.4. METHODS BASED ON QUADRATIC EQUATIONS 51

$\mathcal{G}$ be the reduced Gröbner basis of $J(t, \mathbf{r})$ w.r.t the lexicographic order induced by $U_1 < \cdots < U_n < V_1 < \cdots < V_n$. Then there exists a set $\mathcal{G}' := \{f_i(U_i)\}_{1 \le i \le n} \subset \mathcal{G}$, f_i are square-free polynomials. Moreover, $\mathcal{G}'$ is the reduced Gröbner basis of $I := J(t, \mathbf{r}) \cap \mathbb{F}_q[U_1, \ldots, U_n]$ w.r.t the lexicographic order induced by $U_1 < \cdots < U_n$.

Proof. Since $J(t, \mathbf{r})$ is also zero-dimensional in this case due to Theorem 1.4.42 and t is the smallest such that $J(t, \mathbf{r})$ has a solution, we analogously to the previous proof have that for $I := J(t, \mathbf{r}) \cap \mathbb{F}_q[U_1, \ldots, U_n] : I \cap \mathbb{F}_q[U_i] = \langle f_i \rangle, i = 1, \ldots, n$, where f_i are square-free polynomials.

Now let us show that the above polynomials $f_i, i = 1, \ldots, n$ are in the reduced Gröbner basis of I w.r.t lexicographic order with $U_1 < \cdots < U_n$. Indeed, due to Theorem 1.2.7 we have that the Gröbner basis of I contains polynomials $\tilde{f}_i \in \mathbb{F}_q[U_1, \ldots, U_i]$ such that $lt(\tilde{f}_i) = U_i^{m_i}$. Obviously, $m_i \le \deg(f_i)$. On the other hand if $m_i < \deg(f_i)$ for some i, then not all the elements of $V(I)$ will be recovered. So $m_i = \deg(f_i), i = 1, \ldots, n$. Denote $\tilde{\mathcal{G}} := \{\tilde{f}_1, \ldots, \tilde{f}_n\}$. We have $L(\tilde{\mathcal{G}}) = L(I) = \langle U_1^{m_1}, \ldots, U_n^{m_n} \rangle = L(\mathcal{G}')$. So $\mathcal{G}'$ is also a Gröbner basis. Since $\mathcal{G}'$ is reduced, we have that $\mathcal{G}'$ is the reduced Gröbner basis of I. Since lexicographic order is an elimination order, we have that $\mathcal{G}' \subset \mathcal{G}$. ◇

Remark 1.4.47
- Experimental evidence suggest that the ideals $J(t, \mathbf{r})$ are radical for all t and $\mathbf{r}$. It seems, though, that proving this fact does not yield any practical advantages over the knowledge that only $J(t, \mathbf{r}) \cap \mathbb{F}_q[U_1, \ldots, U_n]$ is radical, since we are interested mainly in U-variables. However for the minimum distance finding the above techniques do not work anymore, since $I(t, \mathcal{U}) + J(\mathbf{r})$ is not radical in this case.

- We see that in order to find $V(J(t, \mathbf{r}))$ for the case of bounded distance decoding any monomial order will do. In particular fast degree reverse lexicographic order is quite efficient in this case. In order to solve the nearest codeword problem, one has to find the reduced Gröbner basis w.r.t lexicographic order, which is in practice much harder. Nevertheless, we may use the fact that $J(t, \mathbf{r})$ is zero-dimensional: first find a Gröbner basis w.r.t some effective order, e.g. degree reverse lexicographic, and then use FGLM technique to convert it to the one w.r.t lexicographic order. If the number of codewords closest to $\mathbf{r}$ is polynomial in n, then the problem of solving the nearest codeword problem would be essentially the same as finding the reduced Gröbner basis w.r.t any monomial order.

Remark 1.4.48 • We note that in [7] the authors also prove the uniqueness result for cyclic codes. Still they did not succeed to prove that their unique solution has multiplicity one until their paper [9]. Our Theorem 1.4.45 states exactly this for arbitrary linear codes.

• For bounded distance decoding we have that the system $J(t, \mathbf{r})$, for $t = \text{wt}(\mathbf{e})$, has a unique simple solution $(\mathbf{u}, \mathbf{v})$. It is known from Theorem 1.4.45 that $\mathbf{u} = \mathbf{u}(\mathbf{e})$, the unknown syndrome of $\mathbf{e}$, lies in $\mathbb{F}_q^n$. But then via substitution of $\mathbf{u}$ to the system $J(t, \mathbf{r})$ it is not hard to see that $\mathbf{v}$ is also from $\mathbb{F}_q^n$.

We are now ready to formulate the algorithm for bounded distance decoding. In the algorithm it is assumed that the number of errors occurred does not exceed the error correcting capacity of the code as in Theorem 1.4.45.

Algorithm 1.4.49 [Bounded distance decoding]
Input: code C (via check matrix H), MDS basis B, received word $\mathbf{r}$ with the assumption $\mathbf{r} = \mathbf{c} + \mathbf{e}$ for some $\mathbf{c} \in C$ and $\text{wt}(\mathbf{e}) \leq (d(C) - 1)/2$.
Output: The codeword $\mathbf{c}$.

1. Set $i = 1$

2. Find the reduced Gröbner basis G of the ideal $J(i, \mathbf{r})$ with respect to any ordering chosen in advance

3. If $G = \{1\}$, then $i := i + 1$ and go to (2)

4. G is of the form $\{U_j - u_j, V_l - v_l\}_{1 \leq j \leq n, 1 \leq l \leq i}$. Form the vector $\mathbf{u} = (u_1, \ldots, u_n)$ of unknown syndromes of $\mathbf{r}$

5. Compute $\mathbf{e}^T = B^{-1}\mathbf{u}$

6. Return $\mathbf{c} := \mathbf{r} - \mathbf{e}$

Example 1.4.50 Consider the ternary $[4,2,3]$ code with generator matrix $G = (I_2|P)$ and parity matrix $H = (-P^T|I_2)$ with

$$P = \begin{pmatrix} 1 & 1 \\ 1 & -1 \end{pmatrix}.$$

Let $a \in \mathbb{F}_9$ with $a^2 = -1$. Let $\mathbf{b}_i = (1, a^i, a^{2i}, a^{3i})$ and B the 4×4 matrix with $\mathbf{b}_i$ as i-th row. Then

$$\begin{cases} \mathbf{h}_1 &= (1+a)\mathbf{b}_1 + \mathbf{b}_2 + (1-a)\mathbf{b}_3 - \mathbf{b}_4, \\ \mathbf{h}_2 &= -\mathbf{b}_1 \phantom{+ \mathbf{b}_2} - \mathbf{b}_3 + \mathbf{b}_4. \end{cases}$$

1.4. METHODS BASED ON QUADRATIC EQUATIONS 53

Let $\mathbf{r} = (1,1,1,1)$ be a received word. Then $\mathbf{h}_1 \cdot \mathbf{r} = -1$ and $\mathbf{h}_2 \cdot \mathbf{r} = 1$. Hence $J(1, \mathbf{y})$ is generated by the elements

$$\begin{cases} (1+a)U_1 + U_2 + (1-a)U_3 - U_4 + 1, \\ -U_1 - U_3 + U_4 - 1. \end{cases}$$

$$\begin{cases} U_1 V_1 = U_2, \\ U_2 V_1 = U_3, \\ U_3 V_1 = U_4, \\ U_4 V_1 = U_1. \end{cases}$$

Now $U_i = 0$ for some i implies that $U_i = 0$ for all i, which contradicts the first two equations. Hence $U_i \neq 0$ for all i. The second set of equations gives

$$\begin{cases} V_1 = U_2/U_1, \\ U_3 = U_2^2/U_1, \\ U_4 = U_2^3/U_1^2, \\ U_4^4 = U_1^4. \end{cases}$$

Together with the first two equations we get that $(U_1, U_2, U_3, U_4) = (a, -1, -a, 1)$ and $V_1 = a$ is the unique solution. Furthermore $\mathbf{e} = (0, 1, 0, 0)$ is the only vector with $u_i(\mathbf{e}) = U_i$ for all i. Hence $\mathbf{r} - \mathbf{e} = (1, 0, 1, 1)$ is the unique closest codeword.

Example 1.4.51 Let us now demonstrate how extension field is used for correcting codes, which are not MDS. Consider a ternary Golay code with parameters [11,6,5]. The code is 2-error correcting. We have $q = 3$ and $n = 11$. So $m = 3$ is the smallest number, such that $3^m > n$. We have not taken the conventional choice for cyclic codes $m = 5$ which is the smallest degree of an extension such that $\mathbb{F}_{3^m}^*$ has an element of order 11. Let a parity check matrix for this code be (after row reduction):

$$H = \begin{pmatrix} 1 & -1 & -1 & -1 & 1 & 0 & 1 & 0 & 0 & 0 & 0 \\ 0 & 1 & -1 & -1 & -1 & 1 & 0 & 1 & 0 & 0 & 0 \\ -1 & 1 & -1 & 0 & 1 & -1 & 0 & 0 & 1 & 0 & 0 \\ 1 & 1 & 0 & 1 & 1 & 1 & 0 & 0 & 0 & 1 & 0 \\ -1 & -1 & -1 & 1 & 0 & 1 & 0 & 0 & 0 & 0 & 1 \end{pmatrix}.$$

The corresponding MDS matrix $B = (a^{ij})_{0 \leq i,j \leq 10}$, where a is a primitive element of $\mathbb{F}_{27}$ (with $a^3 - a + 1 = 0$). Now, let $\mathbf{c} = 0$ be a codeword, $\mathbf{e} = (0, 1, 0, -1, 0, 0, 0, 0, 0, 0, 0)$ an error vector with two non-zero positions. So the received word then is $\mathbf{r} = \mathbf{e}$. We are working in the ring $\mathbb{F}_{27}[U_1, \ldots, U_{11}, V_1, V_2]$. Let us choose an ordering to be `degrevlex` with $U_{11} > \cdots > U_6 >$

$V_1 > V_2 > U_5 > \cdots > U_1$ (cf. Theorem 1.4.45). The ideal $J(2, \mathbf{r})$ is rather complicated, so we will not list its generators here. Rather we give the reduced Gröbner basis for this ideal:

$$\begin{array}{ccc} & U_1, & \\ U_2 - 1, & U_3 + a^9, & U_4 - 1, \\ U_5 + a^3, & V_2 + a^9, & V_1 + a^4, \\ U_6 + a^{16}, & U_7 + a, & U_8 + a^3, \\ U_9 + a^{22}, & U_{10} - 1, & U_{11} + a. \end{array}$$

One can check that this unique solution indeed gives rise to the error vector $\mathbf{e}$.
We also note that the corresponding system for $t = 1$ does not have solutions.

Example 1.4.52 Let us see an application of Theorem 1.4.45 for the case, when matrix H is already an MDS matrix. In this case the first row of equations in Theorem 1.4.45 is trivial, i.e. just an assignment of values to variables responsible for known syndromes.
As a concrete example let us consider a $[15, 9, 7]$ Reed-Solomon code over $\mathbb{F}_{16}$. It is a 3-error correcting code. Take as parity check matrix for this code the RS matrix $B = (a^{ij})_{0 \leq i,j \leq 14}$, where a is a primitive element of $\mathbb{F}_{16}$, with $a^4 + a + 1 = 0$. Now, let

$$\mathbf{c} = (a^{13}, 0, a^5, a^{14}, a^9, a^{14}, 0, 0, 1, 0, 1, 1, 0, 0, 0)$$

be a codeword,

$$\mathbf{e} = (0, 1, 0, a, a^2, 0, 0, 0, 0, 0, 0, 0, 0, 0, 0)$$

an error vector with three non-zero positions. So the received word then is

$$\mathbf{r} = (a^{13}, 1, a^5, a^7, a^{11}, a^{14}, 0, 0, 1, 0, 1, 1, 0, 0, 0).$$

We are working in the ring $\mathbb{F}_{16}[U_1, \ldots, U_{15}, V_1, V_2, V_3]$. The ordering is degree-reverse lexicographic with $U_{15} > \cdots > U_7 > V_1 > V_2 > V_3 > U_6 > \cdots > U_1$. The ideal $J(3, \mathbf{r})$ is then

$$\begin{array}{ccc} U_1 + a^{10}, & U_2 + a^{13}, & U_3 + a^3 \\ U_4 + a^5, & U_5 + a^5, & U_6 + a^{12}, \end{array}$$
$$V_1 U_i + V_2 U_{i+1} + V_3 U_{i+2} + U_{i+3}, \text{ for } i = 1, \ldots, 15,$$

(1.11)

1.4. METHODS BASED ON QUADRATIC EQUATIONS

where the index i is taken modulo 15. The reduced Gröbner basis for this ideal is:
$$\begin{array}{lll} U_1 + a^{10}, & U_2 + a^{13}, & U_3 + a^3, \\ U_4 + a^5, & U_5 + a^5, & U_6 + a^{12}, \\ V_3 + a^{14}, & V_2 + a^{11}, & V_1 + a^8, \\ U_7 + 1, & U_8 + 1, & U_9 + 1, \\ U_{10} + a, & U_{11} + a^9, & U_{12} + a^{12}, \\ U_{13} + a, & U_{14}, & U_{15} + a^{14}. \end{array}$$

One can check that this unique solution indeed gives rise to the error vector **e**. We also note that the corresponding systems for $t = 1$ and $t = 2$ do not have solutions.

A more deep consideration of the situation in the previous example is presented in the next subsection.

Example 1.4.53 Let us consider a $[4, 2, 3]$ 1-error correcting code C over $\mathbb{F}_9$ with a parity check matrix $H = (a^{ij})_{i=0,1; 0 \leq j \leq 3}$, where a is a primitive element of $\mathbb{F}_9$ (with $a^2 + a - 1 = 0$). The corresponding MDS matrix is then $B = (a^{ij})_{0 \leq i,j \leq 3}$. Consider a received word $\mathbf{r} = (1, 0, 1, 0)$. We will see that this received word has 6 closest codewords in our code C at a distance 2. The ideal $J(2, \mathbf{r})$ is

$$-U_2 + a^3 U_1 + a^7, \quad U_2 + a^6 U_1 + a^5, \quad V_1 U_1 + V_2 U_2 - U_3,$$
$$V_1 U_2 + V_2 U_3 - U_4, \quad V_1 U_3 + V_2 U_4 + a^6 U_4 + U_3 + a U_2 + a^6 U_1,$$
$$V_1 U_4 + a^2 V_2 U_4 - V_2 U_3 + a^5 V_2 U_2 + a^2 V_2 U_1 - U_4 + U_3 + U_2 + U_1$$

The reduced Gröbner basis for the ideal $J(2, \mathbf{r})$ w.r.t lexicographic ordering with $U_1 < \cdots < U_4 < V_1 < V_2$ is

$$\begin{array}{l} U_1 + 1, \quad U_2 + a^7, \quad U_3^6 - U_3^5 + U_3^2 - U_3, \\ U_4 + a U_3^5 + a^7 U_3^4 + a^7 U_3^3 - U_3^2 + a^5, \\ V_2 + a^3 U_3^5 + a^7 U_3^4 + a^2 U_3^3 + a U_3^2 + a U_3 + a^7, \\ V_1 + a^6 U_3^5 + a^2 U_3^4 + a^5 U_3^3 - U_3^2 + a^2. \end{array}$$

We are only interested in the first four equations. The first two yield $U_1 = -1, U_2 = a^3$. The third equation yields solutions for U_3:

$$U_3^6 - U_3^5 + U_3^2 - U_3 = U_3(U_3 - 1)(U_3 - a)(U_3 - a^3)(U_3 - a^5)(U_3 - a^7).$$

So we have six solutions for U_3, namely $0, 1, a, a^3, a^5, a^7$. From the forth equation we obtain corresponding values of U_4. They are $a, a^6, a^6, 1, a^6, -1$.

After multiplying these syndrome vectors by the inverse of B we have the six closest codewords:

$$\mathbf{c}_1 = (0,0,0,0), \quad \mathbf{c}_2 = (a, a^3, 1, 0), \quad \mathbf{c}_3 = (1, a, 1, a^2),$$
$$\mathbf{c}_4 = (1, 0, a^6, a^5), \quad \mathbf{c}_5 = (a^2, 0, 1, a^7), \quad \mathbf{c}_6 = (1, a^2, a^7, 0).$$

The following results give an opportunity to rewrite our system from Definition 1.4.37 in a more natural way.

Theorem 1.4.54 *The following systems of equations are equivalent*

$$\begin{cases} \sum_{j=1}^n U_{ij} V_j = U_{i,t+1} & \text{for all } i = 1, \ldots, n, \\ \sum_{j=1}^n a_{ij} U_j = s_i(\mathbf{r}) & \text{for all } i = 1, \ldots, n-k. \end{cases}$$

and

$$\begin{cases} X_i Y_i = 0 & \text{for all } i = 1, \ldots, n, \\ \sum_{j=1}^n h_{ij} X_j = s_i(\mathbf{r}) & \text{for all } i = 1, \ldots, n-k, \\ \sum_{j=1}^n c_{j,+1} Y_j = -1, \\ \sum_{j=1}^n c_{ji} Y_j = 0 & \text{for all } i = t+2, \ldots, n. \end{cases}$$

Proof. Let $U = (U_1, \ldots, U_n)^T$ and $X = (X_1, \ldots, X_n)^T$. Let $Y = (Y_1, \ldots, Y_n)^T$ and $V = (V_1, \ldots, V_t, -1, 0, \ldots, 0)^T$ a vector of length n. If $U = BX$ and $Y = B^T V$. Now $X * Y = (X_1 Y_1, \ldots, X_n Y_n)^T$ and

$$B \cdot (X * Y) = BD(X)B^T V = \mathcal{U}V,$$

since $\mathcal{U}(X) = BD(X)B^T$ by Proposition 1.4.6. B is invertible, hence the following statements are equivalent

$$\begin{array}{rcl}
\sum_{j=1}^n U_{ij} V_j & = & U_{it+1} \quad \text{for all } i = 1, \ldots, n, \\
\mathcal{U}V & = & 0, \\
B(X * Y) & = & 0, \\
X * Y & = & 0, \\
X_i Y_i & = & 0 \quad \text{for all } i = 1, \ldots, n.
\end{array}$$

Now $HX = ABX = AU = \mathbf{s}(\mathbf{r})$. So the following statements are equivalent

$$\begin{array}{ll}
\sum_{j=1}^n a_{ij} U_j = s_i(\mathbf{r}) & \text{for all } i = 1, \ldots, n-k, \\
\sum_{j=1}^n h_{ij} X_j = s_i(\mathbf{r}) & \text{for all } i = 1, \ldots, n-k.
\end{array}$$

Note that the first equivalence holds under $Y = B^T V$. Hence $B^{-T} Y = V$, $V_{t+1} = -1$, and $V_j = 0$ for all $i > t+1$. The entries of B^{-1} are c_{ij}. Therefore the statement

$$V_{t+1} = -1 \text{ and } V_i = 0 \text{ for all } i > t+1,$$

1.4. METHODS BASED ON QUADRATIC EQUATIONS

is equivalent with

$$\sum_{j=1}^{n} c_{j,t+1} Y_j = -1 \text{ and } \sum_{j=1}^{n} c_{ji} Y_j = 0 \text{ for all } i > t+1.$$

◇

Corollary 1.4.55 *The two systems from Theorem 1.4.54 are equivalent with the following system:*
$$\begin{cases} HX = \mathbf{s}(\mathbf{r}), \\ X * Y = 0, \\ \hat{H}_t Y = \hat{H}_t \mathbf{b}_{t+1}^T, \end{cases}$$
where $X = (X_1, \ldots, X_n)^T$, $Y = (Y_1, \ldots, Y_n)^T$, and $\hat{H}_t$ is a parity check matrix of the code with generator matrix B_t.

Proof. We proceed as in the proof of Theorem 1.4.54. Note that $Y = B^T V$ is equivalent to $Y = B_t^T V' + \mathbf{b}_{t+1}^T$, where $V' = (V_1, \ldots, V_t)^T$. This in turn can be written as $Y - \mathbf{b}_{t+1}^T = B_t^T V$. For a parity check matrix $\hat{H}_t$ we have $B_t \hat{H}_t^T = \hat{H}_t B_t^T = 0$. Therefore, we have $\hat{H}_t(Y - \mathbf{b}_{t+1}^T) = 0$. This finishes the proof. ◇

Remark 1.4.56 Now we see that earlier the U-variables were responsible for an unknown syndrome, now X-variables yield an error vector directly. Also note that $\hat{H}_t \mathbf{b}_{t+1}^T$ in Corollary 1.4.55 is actually the syndrome of $\mathbf{b}_{t+1}$ with respect to the code with generator matrix B_t. We use this more coding-theoretic formulation in the next section.

1.4.4 Generalized Newton identities for arbitrary linear codes

In this subsection we show how to obtain generalized Newton identities (GNI) for an arbitrary linear code identical to those that exist for cyclic codes as in Section 1.3.2. Namely, we show that by suitably choosing an MDS matrix B we can obtain the ideal $I(t, \mathcal{U}, V)$ in the form as in (1.7). We adopt notation as in Section 1.3.2. So we suppose that $(n, q) = 1$ and a is a primitive n-th root of unity in $\mathbb{F}$. As an MDS matrix we can choose an RS-matrix $B(a)$, cf. Definition 1.4.9. Now following Remark 1.4.10 and the definition of $I(t, \mathcal{U}, V)$ we have that $I(t, \mathcal{U}, V)$ is generated by

$$\sum_{j=1}^{t} U_{((i+j-2) \mod n)+1} V_j - U_{i+t}, i = 1, \ldots, n,$$

So now $I(t,\mathcal{U},\mathcal{V})$ has the form of GNI as in (1.7) up to renumbering of indices. This resemblance in the form of equations is not a coincidence. For a cyclic code we have

Proposition 1.4.57 *For the cyclic code C with a defining set S_C and received vector $\mathbf{r} = \mathbf{c} + \mathbf{e}$ let $s_i, i \in \mathbb{Z}_n$ be the syndromes (both known and unknown) and let σ_j be the elementary symmetric functions. Let $J(t,\mathbf{r})$ be the ideal that corresponds to C and $\mathbf{r}$ constructed w.r.t the RS-matrix $B(a)$. We assume that the number of errors t does not exceed the error capacity of C, so that $J(t,\mathbf{r})$ has a unique solution $(\mathbf{u}(\mathbf{e}), \mathbf{v})$. Then the following hold:*

$$u_i(\mathbf{e}) = s_{i-1}, v_j = -\sigma_{t-j+1}, i = 1, \ldots, n, t = 1, \ldots, t,$$

where $s_0 = s_n$.

Proof. Note that the parity check matrix H of C from (1.1) is a submatrix of $B(a)$. Namely, for $i \in S_C$ the row $\mathbf{h}_i = (1, a^i, \ldots, a^{(n-1)i})$ of H is the $(i+1)$-th row of $B(a)$. Thus for $i \in S_C$ we have

$$u_{i+1}(\mathbf{e}) = \mathbf{b}_{i+1} \cdot \mathbf{e} = \mathbf{h}_i \cdot \mathbf{e} = s_i.$$

Now assign $S_i := u_{i+1}, \sigma_j = -v_{t-j+1} \forall i, j$ in the system (1.9). We see that $(\mathbf{u}, \mathbf{v})$ is now a solution of (1.9) with the latter renumbering of indices. By the result from [41] we have that (1.9) has a unique solution. Thus the claim follows. ⋄

Remark 1.4.58 • The above Proposition shows that it is not necessary to add to (1.9) the field equations or any other equations that represent conjugacy class membership.

• On some computations with GNI for cyclic codes, see Section 1.4.8.

For the case of binary cyclic codes in [7] the authors use Waring function for finding the syndromes via elementary symmetric functions. Note that we can use Waring functions for arbitrary linear codes, as U- and V-variables now are connected via the GNI. We have the following result.

Proposition 1.4.59 *Let $B = B(a)$ be the RS-MDS matrix. Let C be a binary $[n,k]$ code with the parity check matrix $H = AB$. Let $\mathbf{r} = \mathbf{c} + \mathbf{e}$ be the received word. Then $(\mathbf{u}(\mathbf{e}), \mathbf{v})$ is a solution of $J(t,\mathbf{r})$ if and only if $\mathbf{v}$ is a solution of the system*

$$a_{j1} f_n(V_t, \ldots, V_1) + \sum_{l=2}^{n} a_{jl} f_{l-1}(V_t, \ldots, V_1) = s_j(\mathbf{r}), j = 1, \ldots, n-k.$$

1.4. METHODS BASED ON QUADRATIC EQUATIONS

Proof. Let $U_{i+1} = f_i(V_t, \ldots, V_1), 1 \leq i \leq n-1, U_1 = f_n(V_t, \ldots, V_1)$, be the expression of U-variables in terms of V-variables via Waring functions f_i. It exists due to the existence of GNI, for details cf. [87]. Now simply substitute the corresponding expressions in $J(\mathbf{r})$ to obtain the desired equations. ◇

The following result can be found in [9]. Here it is a simple corollary from the preceding material

Theorem 1.4.60 *Let C be a binary cyclic code and $s_1, \ldots, s_n$ the syndromes of the received word $\mathbf{r}$, and let $\sigma_1, \ldots, \sigma_t$ be the elementary symmetric functions. Assume that the number of errors t does not exceed the error capacity of C. Then the system*

$$s_i = f_i(\sigma_1, \ldots, \sigma_t), i \in S_C$$

has a unique solution $(\sigma_1^*, \ldots, \sigma_t^*)$ of multiplicity 1.

1.4.5 Finding the minimum distance

In this subsection we investigate the question, how the techniques from Section 1.4.3 can be adopted to the problem of finding the minimum distance of a given code. The following is more or less a special case of Theorem 1.4.42.

Theorem 1.4.61 *Let B be an MDS matrix with structure constants μ_l^{ij} and linear functions U_{ij}. Let H be a parity check matrix of the code C such that $H = AB$. Let t be the smallest integer such that $J(t,0)$ has a solution $(\mathbf{u}, \mathbf{v})$ with $\mathbf{u} \neq 0$. Then t is the minimum distance of C.*

Proof. Denote $d = d(C)$.
1) Let $\mathbf{c}$ be a codeword of C of weight d. Then $\mathbf{s}(\mathbf{c}) = 0$ and $\mathbf{u} = \mathbf{u}(\mathbf{c}) \neq 0$. So there is a solution $(\mathbf{u}, \mathbf{v})$ of $J(d,0)$ with $\mathbf{u} = \mathbf{u}(\mathbf{c}) \neq 0$ by Lemma 1.4.39.
2) Suppose that t is the smallest positive integer such that $J(t,0)$ has a solution $(\tilde{\mathbf{u}}, \tilde{\mathbf{v}})$ with $\tilde{\mathbf{u}} \neq 0$. Then $t \leq d$ by (1).
There is a unique $\tilde{\mathbf{c}} \in \mathbb{F}_{q^m}^n$ such that $\mathbf{u} = \mathbf{u}(\tilde{\mathbf{c}})$ and $\text{wt}(\tilde{\mathbf{c}}) \leq t$, by Lemma 1.4.40. Note that $\tilde{\mathbf{c}}$ is a codeword of $\tilde{C}$, where $\tilde{C} = C\mathbb{F}_{q^m}$. Now $\mathbf{c} \neq 0$, otherwise $\mathbf{u}' = 0$. Hence $\mathbf{c}'$ is a non-zero codeword of $\tilde{C}$ with $\text{wt}(\tilde{\mathbf{c}}) \leq t \leq d(\tilde{C})$. Furthermore $d = d(\tilde{C})$ by Proposition 1.3.14. So $\text{wt}(\tilde{\mathbf{c}}) = d(\tilde{C}) = d(C) \geq t$. Therefore $t = d$. ◇

Example 1.4.62 Let us find minimum distance of a cyclic $[15, 7]$ binary code with a check matrix:

$$H = \begin{pmatrix} 1 & 1 & 0 & 1 & 0 & 0 & 0 & 1 & 0 & 0 & 0 & 0 & 0 & 0 & 0 \\ 0 & 1 & 1 & 0 & 1 & 0 & 0 & 0 & 1 & 0 & 0 & 0 & 0 & 0 & 0 \\ 0 & 0 & 1 & 1 & 0 & 1 & 0 & 0 & 0 & 1 & 0 & 0 & 0 & 0 & 0 \\ 0 & 0 & 0 & 1 & 1 & 0 & 1 & 0 & 0 & 0 & 1 & 0 & 0 & 0 & 0 \\ 0 & 0 & 0 & 0 & 1 & 1 & 0 & 1 & 0 & 0 & 0 & 1 & 0 & 0 & 0 \\ 0 & 0 & 0 & 0 & 0 & 1 & 1 & 0 & 1 & 0 & 0 & 0 & 1 & 0 & 0 \\ 0 & 0 & 0 & 0 & 0 & 0 & 1 & 1 & 0 & 1 & 0 & 0 & 0 & 1 & 0 \\ 0 & 0 & 0 & 0 & 0 & 0 & 0 & 1 & 1 & 0 & 1 & 0 & 0 & 0 & 1 \end{pmatrix}.$$

By computing the reduced Gröbner basis of $J(t,0)$ w.r.t degree reverse lexicographic order for $t = 1, \ldots, 4$ we see that it always consists of the elements $U_1, \ldots, U_{15}$, so there is no solution $(\mathbf{u}, \mathbf{v})$ with $\mathbf{u} \neq 0$. For $t = 5$ the reduced Gröbner basis is

$U_2, \quad U_3, \quad U_4, \quad U_5, \quad U_7, \quad U_9, \quad U_{10}, \quad U_{13},$
$V_5 U_1, \quad V_3 U_1,$
$V_1 U_1 + U_6, \quad U_6^2 + U_{11} U_1, \quad U_{11} U_6 + U_1^2,$
$V_5 U_6, \quad V_4 U_6 + U_8, \quad V_3 U_6,$
$V_1 U_6 + U_{11}, \quad U_8^2 + U_{15} U_1, \quad U_{12} U_8 + U_{14} U_6,$
$V_5 U_8, \quad V_4 U_8 + V_2 U_6, \quad V_3 U_8,$
$V_2 U_8 + U_{12}, \quad U_{11}^2 + U_6 U_1, \quad U_{12} U_{11} + U_{15} U_8,$
$V_5 U_{11}, \quad V_4 U_{11} + V_1 U_8, \quad V_3 U_{11},$
$V_2 U_{11} + U_{15}, \quad V_1 U_{11} + U_1, \quad U_{12}^2 + U_8 U_1,$
$U_{14} U_{12} + U_{15} U_{11}, \quad V_5 U_{12}, \quad V_4 U_{12} + U_{14},$
$V_3 U_{12}, \quad U_{14}^2 + U_{12} U_1, \quad U_{15} U_{14} + U_8 U_6,$
$V_5 U_{14}, \quad V_4 U_{14} + V_2 U_{12}, \quad V_3 U_{14},$
$V_2 U_{14} + V_4 U_1, \quad U_{15}^2 + U_{14} U_1, \quad V_5 U_{15},$
$V_4 U_{15} + V_1 U_{12}, \quad V_3 U_{15},$
$V_2 U_{15} + V_1 U_{14}, \quad V_1 U_{15} + V_2 U_1,$
$V_4 U_1^2 + U_{11} U_8, \quad V_2 U_1^2 + U_{15} U_6,$
$V_2 U_6 U_1 + U_{15} U_{11}, \quad V_4^2 U_1 + V_2 U_1,$

1.4. METHODS BASED ON QUADRATIC EQUATIONS

$$U_14U_8U_6 + U_{15}U_{12}U_1, \quad U_{15}U_8U_6 + U_{12}U_1^2,$$
$$V_2U_{12}U_6 + U_{14}U_8, \quad V_2^2U_6 + U_{14},$$
$$U_{14}U_{11}U_8 + U_{15}U_{12}U_6,$$
$$U_{15}U_{11}U_8 + U_{12}U_6U_1, \quad V_1U_{14}U_8 + U_{15}U_{12},$$
$$V_1^2U_8 + V_4U_1, \quad V_2^2U_{12} + V_2U_1,$$
$$V_1^2U_{12} + V_2V_4U_1, \quad V_1^2U_{14} + V_2^2U_1,$$
$$V_2^3U_1 + V_1U_8$$

It can be seen that, e.g. $(\mathbf{u}, \mathbf{v})$ with $\mathbf{u} = (1,0,0,0,0,1,0,1,0,0,1,1,0,1,1)$ and $\mathbf{v} = (1,1,0,1,0)$ is a solution of the system $J(5,0)$. So we obtained a desired solution, thus minimum distance is 5. It can also be seen that $\mathbf{u}$ corresponds to a codeword of weight 5, namely $\mathbf{c} = (1,0,0,0,0,0,0,1,0,0,0,1,0,1,1)$. The fact that we obtained a codeword of weight 5 is not a coincidence. See the following proposition.

Proposition 1.4.63 *Let B be an MDS matrix with structure constants μ_l^{ij} and linear functions U_{ij}. Let H be a parity check matrix of the code C such that $H = AB$. Denote by $J_q(t, \mathbf{r})$ an ideal generated by $J(t, \mathbf{r})$ and the field equations $U_i^q - U_i = 0, 1 \leq i \leq n$ and $V_j^q - V_j = 0, 1 \leq j \leq t$. The solutions of $J_q(t,0)$ that are no solution of $J_q(t-1,0)$ correspond one-to one to codewords of C of weight t.*

Proof. 1) Suppose that $\mathbf{c}$ is a codeword of C of weight w. Let $u_i = \mathbf{b}_i \cdot \mathbf{c}$ for all i, then there is unique $\mathbf{v}$ such that $(\mathbf{u}, \mathbf{v})$ is a solution of $J_q(w, 0)$ and not a solution of $J_q(w-1, 0)$.
2) Suppose that $(\mathbf{u}, \mathbf{v})$ is solution over $\mathbb{F}_q$ of $J_q(t, 0)$ that is not a solution of $J_q(t-1, 0)$. Then there is a unique $\mathbf{c} \in \mathbb{F}_q^n$ such that $u_i = \mathbf{b}_i \cdot \mathbf{c}$. Now $s_j(\mathbf{c}) = \sum_{l=1}^n a_{jl} u_l = 0$ for all j. Hence $\mathbf{c}$ is a codeword of C.
Now $(\mathbf{u}, \mathbf{v})$ is not a solution of $J(t-1, 0)$. So $\mathcal{U}_{nv}(\mathbf{c})$ has rank v for all $v \leq t$. But $(\mathbf{u}, \mathbf{v})$ is a solution of $J(t, 0)$. These equations imply that $\mathcal{U}_{n,t+1}(\mathbf{c})$ has rank t. So $\text{wt}(\mathbf{c}) = t$ by Lemma 1.4.13. Hence $\mathbf{c}$ is a codeword of weight t. ◇

Remark 1.4.64 In practice, in order to compute $d(C)$ we look at the smallest t such that the reduced Gröbner basis of $J(t,0)$ w.r.t any monomial order (e.g. degree reverse lexicographic) is not equal to $\{U_1, \ldots, U_n\}$. This has to do with radicality issues of $J(t, \mathbf{r})$ as is outlined in Remark 1.4.47. Note that as direct computations show, in the case $t < d(C)$, $I(t, \mathcal{U}) + J(0)$ is not radical, so arguments similar to those in Theorem 1.4.45 and Theorem 1.4.46 do not apply here.

1.4.6 Generic decoding

In this section we briefly discuss a question of generic decoding of linear codes, i.e. obtaining "closed formulas" for error-values via the known syndrome variables. We provide a quite simple general result and analyze an example.

Theorem 1.4.65 *Let C be a q-ary $[n,k,d]$ linear code. Let $\mathbb{F}_Q$ be an MDS extension (cf. Definition 1.4.9), so that $Q \geq n$ and let $t \leq (d-1)/2$. Consider an ideal $J_q(t)$ in the ring $\mathbb{F}_Q[S_1,\ldots,S_{n-k},X_1,\ldots,X_n,Y_1,\ldots,Y_n]$ generated by*

$$\begin{cases} HX = S, \\ X * Y = 0, \\ \hat{H}_t Y = \hat{H}_t \mathbf{b}_{t+1}^T, \\ S_i^q - S_i = 0, i = 1, \ldots, n-k, \\ X_i^q - X_i = 0, i = 1, \ldots, n, \end{cases}$$

where $X = (X_1,\ldots,X_n)^T, Y = (Y_1,\ldots,Y_n)^T, S = (S_1,\ldots,S_{n-k})^T$, and $\hat{H}_t$ is a parity check matrix of the code with generator matrix B_t. Let $<$ be any order that eliminates Y-variables and restricted to (S, X)-variables is a block order with $S < X$. Then there exist $f_i \in \mathbb{F}_q[S_1,\ldots,S_{n-k}], i = 1,\ldots,n$ such that $X_i - f_i(S_1,\ldots,S_{n-k})$ are in the reduced Gröbner basis of $J_q(t)$ w.r.t. $<$.

Proof. Recall syndrome decoding with coset leaders, Section 1.1. For any syndrome vector $\mathbf{s} \in \mathbb{F}_q^{n-k}$ we have one or several coset leaders in the coset of elements with the same syndrome $\mathbf{s}$. Due to the assumption $t \leq (d-1)/2$ we have that there is always exactly one coset leader for each syndrome $\mathbf{s}$. Let $\mathcal{S}$ be a subset of $\mathbb{F}_q^{n-k}$ that is composed of the syndromes that correspond to error vectors of weight $\leq t$. Then we have a map $\phi : \mathbb{F}_q^{n-k} \supset \mathcal{S} \to \mathbb{F}_q^n$ that maps a syndrome $\mathbf{s}$ to the corresponding coset leader (error vector) $\phi(\mathbf{s})$. Since we are working over the finite field $\mathbb{F}_q$ here, we have that $\phi = (f_1,\ldots,f_n)$ with $f_i \in \mathbb{F}_q[S_1,\ldots,S_{n-k}], i = 1,\ldots,n$. Now the field equations $S_i^q - S_i = 0, i = 1,\ldots,n-k, X_j^q - X_j = 0, j = 1,\ldots,n$ are included to $J_q(t)$, so $I := J_q(t) \cap \mathbb{F}_Q[S_1,\ldots,S_{n-k},X_1,\ldots,X_n]$ is a radical ideal. Therefore $X_i - f_i(S_1,\ldots,S_{n-k}) \in I$. Now the statement easily follows. Note that f_i's are reduced modulo $S_i^q - S_i, i = 1,\ldots,n-k$ and have coefficients from the smaller field $\mathbb{F}_q$, not $\mathbb{F}_Q$. ◇

Example 1.4.66 *Let us consider a random binary $[20, 7, 5]$ code with a*

1.4. METHODS BASED ON QUADRATIC EQUATIONS

parity-check matrix

$$H = \begin{pmatrix} 1 & 1 & 1 & 1 & 0 & 1 & 0 & 1 & 0 & 0 & 0 & 0 & 0 & 0 & 0 & 0 & 0 & 0 & 0 \\ 1 & 0 & 0 & 1 & 1 & 0 & 0 & 0 & 1 & 0 & 0 & 0 & 0 & 0 & 0 & 0 & 0 & 0 & 0 \\ 0 & 0 & 1 & 1 & 0 & 1 & 1 & 0 & 0 & 1 & 0 & 0 & 0 & 0 & 0 & 0 & 0 & 0 & 0 \\ 0 & 1 & 0 & 0 & 0 & 0 & 1 & 0 & 0 & 0 & 1 & 0 & 0 & 0 & 0 & 0 & 0 & 0 & 0 \\ 0 & 0 & 0 & 0 & 1 & 0 & 1 & 0 & 0 & 0 & 0 & 1 & 0 & 0 & 0 & 0 & 0 & 0 & 0 \\ 1 & 1 & 0 & 0 & 1 & 0 & 1 & 0 & 0 & 0 & 0 & 0 & 1 & 0 & 0 & 0 & 0 & 0 & 0 \\ 1 & 0 & 1 & 0 & 1 & 0 & 1 & 0 & 0 & 0 & 0 & 0 & 0 & 1 & 0 & 0 & 0 & 0 & 0 \\ 1 & 1 & 0 & 0 & 0 & 1 & 0 & 0 & 0 & 0 & 0 & 0 & 0 & 0 & 1 & 0 & 0 & 0 & 0 \\ 0 & 0 & 1 & 1 & 1 & 0 & 1 & 0 & 0 & 0 & 0 & 0 & 0 & 0 & 0 & 1 & 0 & 0 & 0 \\ 1 & 0 & 1 & 0 & 0 & 1 & 0 & 0 & 0 & 0 & 0 & 0 & 0 & 0 & 0 & 0 & 1 & 0 & 0 \\ 1 & 1 & 1 & 1 & 1 & 1 & 0 & 0 & 0 & 0 & 0 & 0 & 0 & 0 & 0 & 0 & 0 & 1 & 0 \\ 0 & 1 & 1 & 1 & 0 & 1 & 0 & 0 & 0 & 0 & 0 & 0 & 0 & 0 & 0 & 0 & 0 & 0 & 1 \\ 0 & 1 & 1 & 1 & 1 & 0 & 1 & 0 & 0 & 0 & 0 & 0 & 0 & 0 & 0 & 0 & 0 & 0 & 1 \end{pmatrix}.$$

So we have $q = 2, Q = 32, t = 2$. Let $<$ be a block order with $S < X < Y$, where an order for every block is the degree reverse lexicographic. Computing the reduced Gröbner basis of $J_2(2)$ w.r.t $<$ yields the desired "closed formulas" for the error values X. For example, the formula for X_1 is

$$X_1 = S_5S_4S_3S_2S_1 + S_7S_6S_3S_2 + S_7S_6S_2S_1 + S_8S_5S_2S_1 +$$
$$+ S_7S_5S_2S_1 + S_7S_4S_2S_1 + S_5S_4S_2S_1 + S_5S_3S_2S_1 +$$
$$+ S_8S_5S_4 + S_7S_6S_3 + S_8S_5S_3 + S_6S_5S_3 + S_7S_4S_3 +$$
$$+ S_5S_4S_3 + S_8S_6S_2 + S_6S_5S_2 + S_7S_4S_2 + S_6S_4S_2 +$$
$$+ S_8S_3S_2 + S_6S_3S_2 + S_7S_6S_1 + S_8S_5S_1 + S_6S_5S_1 +$$
$$+ S_7S_4S_1 + S_5S_4S_1 + S_8S_2S_1 + S_6S_2S_1 + S_5S_2S_1.$$

Total degree of f_i's is 4–5. And the number of monomials not including the X_i is 22–58, the mean being 40. So we see that although $n - k = 13$, total degree does not go higher than 5 and the average number of monomials, 40, is much less that an expected number of monomials in a random polynomial in variables $S_1, \ldots, S_{13}$, which is 2^{12}. It is also noticeable that the variables S_{11}, S_{12}, S_{13} appear only in the formulas for X_{18}, X_{19}, X_{20} resp. as linear terms.

1.4.7 Complexity issues

In this subsection we investigate possibilities to estimate the complexity of decoding method developed in Section 1.4.3. Estimating complexity for the problem of solving systems of algebraic equations is hard in general. It is even harder to have estimates for particular systems that would yield practically applicable figures. As to our knowledge only in papers [19, 20] the issue of complexity was addressed explicitly. There the method proposed by authors is shown to have complexity similar to the one of syndrome decoding. So far different authors did not go far in a sense of complexity estimates: only some experimental data is available for solving corresponding systems of equations [31, 7, 6]. Here we give some ideas on how one can obtain an upper bound of complexity for solving systems that appear in our quadratic system method. In particular, we apply some weak form of extended linearization and estimate complexity via Gaussian elimination. This subsection actually presents and is based on several assumptions/conjectures. Solving and clarifying these seems to be quite challenging and involved. So here we present only preliminary analysis.

Analyzing the Macaulay matrix of the system

In order to get complexity estimates we need first to analyze the Macaulay matrix of our system. We will concentrate on the equivalent representation from Corollary 1.4.55, since then it is possible to give a coding-theoretic interpretation for arguments. Form Corollary 1.4.55 we know that $J(i, \mathbf{r})$ is equivalent to

$$HX = \mathbf{s}, \tag{1.12}$$
$$X_i Y_i = 0, i = 1, \ldots, n, \tag{1.13}$$
$$\hat{H}_t Y = \hat{\mathbf{s}}_t, \tag{1.14}$$

where H is a parity check matrix of the code C, $\mathbf{s}$ is the received word $\mathbf{r}$ syndrome, $X = (X_1, \ldots, X_n)^T$ and $Y = (Y_1, \ldots, Y_n)^T$ are new variables, $\hat{H}_t$ is a parity check matrix of a code with the generator matrix B_t, $\hat{\mathbf{s}}_t$ is the syndrome of the vector $\mathbf{b}_{t+1}$ with respect to $\hat{H}_t$ and t is the number of errors. We denote also by B_t the code with the generator matrix B_t. Now note that the matrix H and the vector $\mathbf{s}$ are defined over the base field of constants $\mathbb{F}_q$ whereas $\hat{H}_t$ and $\hat{\mathbf{s}}_t$ are defined over the extension field $\mathbb{F}_{q^m} = \mathbb{F}_Q$, which we need for the construction of an MDS basis.

We want to rewrite the system (1.12)-(1.14) so that it contains only quadratic equations. First, we perform Gaussian elimination for the system (1.12) and express some $n - k$ error variables X via k others, or we may suppress this step, by assuming that the parity check matrix H is given in the systematic form. Next we perform Gaussian elimination for the system (1.14) and express some $n - t$ Y-variables via t others. We then substitute expressions for the expressed $n - k$ X-variables and $n - t$ Y-variables in the system (1.13). Without loss of generality we may assume that this new system Sys now has variables $X_1, \ldots, X_k, Y_1, \ldots, Y_t$ and consists of n quadratic equations. The monomials that appear in the system are $X_1, \ldots, X_k, Y_1, \ldots, Y_t$

1.4. METHODS BASED ON QUADRATIC EQUATIONS

and $X_i Y_j, 1 \leq i \leq k, 1 \leq j \leq t$. Thus the total number of monomials appearing in the system is $kt + k + t = (k+1)(t+1) - 1$. Next, consider the *Macaulay matrix* A which rows are indexed by the equations of *Sys* and which columns are indexed by the monomials. The entry $A_{f,m}$ is equal to coefficient of the monomial m in the equation (polynomial) f. In other words, if $A_{f,m} = a$, then f includes the term am.

We now conjecture the following

Conjecture 1.4.67 *Let C be the random $[n, k]$ code over $\mathbb{F}_q$, defined e.g. by a random full-rank $(n - k) \times n$ parity check matrix H and let $\mathbf{e}$ be a random error vector over $\mathbb{F}_q$ of weight t. Denote by A the $n \times (kt + k + t)$ Macaulay matrix of the system Sys as above. Then the probability of the fact that A has full-rank tends to 1 as n tends to infinity, i.e.*

$$pr\big(rank(A) = \min\{n, kt + k + t\}\big) \to 1, n \to \infty.$$

We next provide some arguments that support the conjecture. We would like to prove that $\mathrm{pr}\big(\mathrm{rank}(A) < \min\{n, kt + k + t\}\big) \to 0, n \to \infty$. For the proof one may start with the following. First, we consider the case $n \leq kt+k+t$ and show that degeneracy of A is equivalent to an inclusion (1.16) of a restriction of a coset defined by the error vector to an orthogonal complement of a restricted MDS code. Then due to randomness of the code C and the error vector $\mathbf{e}$ it is reasonable to conjecture that such an inclusion is small. The case $n > kt + k + t$ may be then done analogously by considering every full square submatrix of A.

So first consider the case $n \leq kt + k + t$. Here $\mathrm{rank}(A)$ becomes the row rank of A. If A is not full-rank, then this means that the rows of A are linearly dependent. So for $1 \leq i \leq n$ there exist $\alpha_i \in \mathbb{F}_Q$ not all equal to zero such that $\sum_{i=1}^n X_i Y_i = 0$, where X's satisfy additionally equations (1.12) and Y's satisfy (1.14). This means that for every $x \in \mathbf{e} + C$ and every $y \in \mathbf{b}_{t+1} + B_t$ holds $\sum_{i=1}^n x_i y_i = 0$. Let $l = |\{i | \alpha_i \neq 0\}|$. Assume for now that $\alpha_1, \ldots, \alpha_l$ are non-zero. After the linear transformation defined by $Y_i \mapsto \alpha_i Y_i$ for $1 \leq i \leq n$, we have that $\sum_{i=1}^l X_i Y_i = 0$. This, in turn, means that

$$\sum_{i=1}^l x_i y_i = 0 \qquad (1.15)$$

for all $x \in \mathbf{e}_l + C_l$ and all $y \in \widetilde{\mathbf{b}_{t+1}} + \widetilde{B}_t$. Here $\mathbf{e}_l$ and C_l are the vector $\mathbf{e}$ and the code C resp. restricted to the first l positions. Also $\widetilde{\mathbf{b}_{t+1}}$ and $\widetilde{B}_t$ are $\mathbf{b}_{t+1}$ and B_t resp. after applying the above linear transformation. So that e.g. $\widetilde{B}_t$

is the code equivalent to the code B_t restricted to the first l positions.

The equation (1.15) implies that $\mathbf{e}_l + C_l \subseteq (\widetilde{\mathbf{b}_{t+1} + \widetilde{B}_t})^\perp$. Next we note that $(\widetilde{\mathbf{b}_{t+1} + \widetilde{B}_t})^\perp = (\widetilde{B_{t+1}})^\perp$. Indeed $\mathbf{b}_{t+1} + \widetilde{B}_t \subseteq B_{t+1}$, so $(\widetilde{B_{t+1}})^\perp \subseteq (\widetilde{\mathbf{b}_{t+1} + \tilde{B}_t})^\perp =: N$. If $a \in N$, then in particular $\langle a, \widetilde{\mathbf{b}_{t+1}}\rangle = 0$. Also in particular, for every $1 \leq i \leq t$: $\langle a, \widetilde{\mathbf{b}_{t+1}}\rangle + \langle a, \widetilde{\mathbf{b}_i}\rangle = 0$ and thus $\langle a, \widetilde{\mathbf{b}_i}\rangle = 0$. Now it is obvious that $N \subseteq (\widetilde{B_{t+1}})^\perp$. Summing up, we have

$$\mathbf{e}_l + C_l \subseteq (\widetilde{B_{t+1}})^\perp. \tag{1.16}$$

Let us study the code $(\widetilde{B_{t+1}})^\perp$ closely. The code B_{t+1} is MDS with the parameters $[n, t+1, n-t]$.

Proposition 1.4.68 *Let C be an $[n, k, d]$ code. Then the punctured code C_I (where punctured positions are $\{1, \ldots, n\} \setminus I$) is an $[|I|, k_I, d_I]$ code with*

$$d - n + |I| \leq d_I \leq d, k - n + |I| \leq k_I \leq k.$$

If moreover $n - |I| < d$, then $k_I = k$.

From Proposition 1.4.68 we have that the code $\widetilde{B_{t+1}}$ has parameters $[l, k_l, d_l]$, where $t + 1 - n + l \leq k_l \leq t + 1$ and $l - t \leq d_l \leq n - t$. From the same Proposition it also follows that if $n - l < n - t$, then $k_l = t + 1$.

Lemma 1.4.69 *Any t rows of A are linearly independent.*

Proof. Let the matrix H after the Gaussian elimination be of the form $(H' | I_{n-k})$, where H' is an $(n-k) \times k$ matrix with entries $\alpha'_{il}, i = 1, \ldots, n-k, l = 1, \ldots, k$. Consider an expression of variables $X_{k+1}, \ldots, X_n$ via $X_1, \ldots, X_k$ from the parity check equations (1.12):

$$X_i = -\sum_{l=1}^{k} \alpha'_{il} X_l + \alpha_i, i = k+1, \ldots, n.$$

Here $\alpha_{il} \in \mathbb{F}_q$ and $\alpha_i \in \mathbb{F}_q$ are corresponding values for the right hand side of (1.12) after Gaussian elimination. if we assume that H is already in the row-echelon form, then $\alpha_i = s_i$ for all $i = 1, \ldots, n-k$, where s_i are components of s. Now assign $\alpha_{il} = -\alpha'_{il}$ for all valid i and l.

Next, let the matrix $\tilde{H}_t$ after the Gaussian elimination be of the form $(M | I_{n-t})$, where M is an $(n-t) \times t$ matrix with entries $\beta'_{jl}, j = 1, \ldots, n-t, l = 1, \ldots, t$. Let an expression of variables $Y_{t+1}, \ldots, Y_n$ via $Y_1, \ldots, Y_t$ be

$$Y_j = -\sum_{l=1}^{t} \beta'_{jl} Y_l + \beta_i, j = t+1, \ldots, n.$$

1.4. METHODS BASED ON QUADRATIC EQUATIONS 67

Here β_i are components of the vector $\hat{s}$ after the Gaussian elimination. Now assign $\beta_{jl} = -\beta'_{jl}$ for all valid j and l.

Some remarks on the matrix M. We claim that every t rows thereof are linearly independent. Indeed, the matrix $(M|I_{n-t})$ is a generator matrix of the code C_M equivalent to $B_t^\perp$. Since B_t is MDS with the parameters $[n, t, n-t+1]$, then C_M is also MDS with the parameters $[n, n-t, t+1]$, cf. [93]. If some t rows of M would be linearly dependent that would imply the existence of a codeword in C_M of weight $\leq t$, a contradiction.

It is convenient to consider submatrices of M formed by certain rows of M. Suppose that $k > t$. Consider first the matrix $A[1..t]$ composed of the first t rows of M (we denote by $A[1..p]$ the matrix $A_{p,kt+k+t}$ as per Definition 1.4.12). The corresponding equations are $X_i Y_i = 0, i = 1 \ldots, t$. So that in $A[1..t]$ the i-th row has exactly one 1 corresponding to $X_i Y_i$. Next, consider $A[t+1..k]$ composed of the next $k-t$ rows. The corresponding equations are

$$X_i(\sum_{l=1}^{t} \beta_{il} Y_l + \beta_i) = 0, i = t+1, \ldots, k.$$

So in the i-th equation ($i = t+1, \ldots, k$) the monomial $X_i Y_h$ has a coefficient β_{ih}. A coefficient for X_i is β_i.

Move on to $A[k+1..n]$. The corresponding equations are

$$(\sum_{l=1}^{k} \alpha_{il} X_l + \alpha_i)(\sum_{l=1}^{t} \beta_{il} Y_l + \beta_i) = 0, i = k+1, \ldots, n.$$

So in the i-th equation ($i = k+1, \ldots, n$) the monomial $X_l Y_h$ has a coefficient $\alpha_{il} \beta_{ih}$. A coefficient for X_l is $\alpha_{il} \beta_i$. A coefficient for Y_h is $\alpha_i \beta_{ih}$.

Summing up the above, we may say that the matrix $A[k+1..n]$ has as a submatrix the matrix obtained from H' by expanding every non-zero entry in a row with index i to the $(k-t+i)$-th row of M multiplied by this entry. A zero entry is expanded with t zeroes in a row. Analogously, $A[t+1..k]$ has as a submatrix the matrix obtained from I_{k-t} by expanding every entry equal to 1 in a row with index i to the i-th row of M. Consider $A[t+1..n]$, for the moment let us collapse the expansions back to initial entries. Then if we choose a submatrix of some t rows, then we see that every column of such a submatrix has $\leq t$ non-zero entries, and thus in the expansion will have a block of $\leq t$ rows of M. Due to the above remarks on M, we have that the initial t rows are linearly independent. The first t rows of A do not affect this situation and thus any t rows of A are linearly independent. The case $k \leq t$ is done analogously. ◇

We have that $l > t$ due to Lemma 1.4.69. So $k_l = t+1$. Singleton bound then yields $d_l = l - t$. So the code $\widetilde{B_{t+1}}$ is MDS with the parameters

$[l,t+1,l-t]$ and thus $(\widetilde{B_{t+1}})^\perp$ is also MDS with the parameters $[l, l-t-1, t+2]$.

Now, from (1.16) we have in particular that $\mathbf{e}_l \in (\widetilde{B_{t+1}})^\perp$. Note that $\text{wt}(\mathbf{e}) = t$, but $d((\widetilde{B_{t+1}})^\perp) = t + 2$, so $\mathbf{e}_l \in (\widetilde{B_{t+1}})^\perp \iff \mathbf{e}_l = 0$.

So, now we know that $\mathbf{e}_l = 0$, then it follows from (1.16) that $C_l \subseteq (\widetilde{B_{t+1}})^\perp$. Considering that C_l is defined over $\mathbb{F}_q$ we have that $C_l \subseteq (\widetilde{B_{t+1}})^\perp \cap \mathbb{F}_q^l =: \mathcal{B}$. Now $\dim_{\mathbb{F}_q} \mathcal{B} \leq \dim_{\mathbb{F}_Q}(\widetilde{B_{t+1}})^\perp = l-t-1$. Next $\dim_{\mathbb{F}_q} C_l = \min\{k, l\}$ with high probability. Due to the fact that $\dim_{\mathbb{F}_q} C_l \leq l - t - 1$ we conclude that $\dim_{\mathbb{F}_q} C_l = k$.

So now we see that for degeneracy of A we need to have that the "random" code C_l is a subcode of the "fixed" code $\mathcal{B}$. One can show that indeed such an inclusion has asymptotically negligible probability. The problem is, though, that we have to run the argument above for all possible support sets $L = \{i | \alpha_i \neq 0\}$ and moreover one has to handle all possible equivalent codes $\widetilde{B_{t+1}}$, which are defined though $\alpha_i \in \mathbb{F}_Q^*, i \in L$. In order to make the resulting probability low, one has to provide more refined analysis of the subfield subcodes $\mathcal{B}$, which seems a very challenging task indeed: recall that e.g. cyclic codes are subfield subcodes of certain MDS codes, and here we are dealing with a more general situation.

The second case is $n > kt + k + t$. Now $\text{rank}(A)$ is the column rank. We want to show that $\text{pr}(\text{rank}(A) < kt + k + t) \to 0, n \to \infty$. The fact that $\text{rank}(A) < kt + k + t$ implies that for every choice $I = \{i_1, \ldots, i_{kt+k+t}\} \subset \{1, \ldots, n\}$ of rows of A, the $(kt + k + t) \times (kt + k + t)$ submatrix A_I of A composed of those rows is degenerate. Let $p(n)$ denote the highest probability that A_I is degenerate. Then

$$\text{pr}(\text{rank}(A) < kt + k + t) \leq p(n)^{\binom{n}{kt+k+t}}.$$

Let $I = \{i_1, \ldots, i_{kt+k+t}\} \subset \{1, \ldots, n\}$ be the chosen rows. If A_I is degenerate, then for all $i \in I$ there exist $\alpha_i \in \mathbb{F}_Q$ not all equal to zero such that $\sum_{i \in I} x_i y_i = 0$ for all $x \in \mathbf{e}_I + C_I$ and all $y \in (\mathbf{b}_t)_I + (B_t)_I$. Let $I' = \{i | \alpha_i \neq 0\}$ and $l' = |I'|$. After a linear transformation defined by $Y_i \mapsto \alpha_i Y_i$ for $i \in I$, we have that $\sum_{i \in I'} x_i y_i = 0$, for all $x \in \mathbf{e}_{I'} + C_{I'}$ and all $y \in \widetilde{\mathbf{b}_{t+1}} + \widetilde{B_t}$. Here the notation is analogous to the one for the case $n \leq kt + k + t$, but the restriction is to the positions indexed by I'. So we end up with the situation as in the case of $n \leq kt + k + t$ with the difference that $l = l'$. By the same argument as above it should be possible to show that $p(n) \to 0, n \to \infty$, so $\text{pr}(\text{rank}(A) < kt + k + t) \to 0, n \to \infty$.

Conjecture 1.4.67 can be supported by encouraging experimental results already for small parameters. In the following table we present the results for the lengths $14 = 4 \cdot 2 + 4 + 2$ and $27 = 6 \cdot 3 + 6 + 3$. In the upper row the code parameters are given as (n, k, t), the middle row shows $kt + k + t$, and

1.4. METHODS BASED ON QUADRATIC EQUATIONS 69

the lower row indicates the number of full-rank cases/non full-rank cases out of 100 random instances.

(14,4,2)	(14,4,1)	(14,4,3)	(14,2,4)	(14,1,4)	(14,3,4)
14	9	21	14	9	21
39/61	99/1	55/45	10/90	70/30	39/61
(27,6,3)	(27,5,3)	(27,7,3)	(27,3,6)	(27,3,5)	(27,3,7)
27	23	31	27	23	31
69/31	100/0	86/14	7/93	90/10	11/89

We can achieve better results if we forbid H' with zero rows (or intentionally put ones in the zero rows). Note that as k grows the situation with all-zero-rows becomes improbable.

(14,4,2)	(14,4,1)	(14,4,3)	(14,2,4)	(14,1,4)	(14,3,4)
14	9	21	14	9	21
74/26	99/1	100/0	72/28	100/0	97/3
(27,6,3)	(27,5,3)	(27,7,3)	(27,3,6)	(27,3,5)	(27,3,7)
27	23	31	27	23	31
95/5	100/0	100/0	81/19	99/1	98/2

Estimating complexity

Having conjectured that the Macaulay matrix of our system is almost always full-rank and making some additional assumptions, we are able to give some upper bounds on complexity.

Definition 1.4.70 Given a decoding algorithm for a code C of rate R over $\mathbb{F}_q$ of complexity $Compl(C)$, the *complexity coefficient* $CC(R)$ is defined as

$$\frac{1}{n}\log_q(Compl(C)),$$

so that

$$Compl(C) = q^{nCC(R)}.$$

Our benchmarks are going to be the exhaustive search together with syndrome decoding and the covering set decoding algorithm, see Section 1.1. A graph for complexity coefficient of our benchmarks is depicted on Figure 1.4.1. In all these graphs information rate R is on the x-axis and the complexity coefficient is on the y-axis.

Let us assume that $n \geq kt + k + t$. Having this assumption we may apply a well-known technique of *linearization*. Namely, note that since we assumed $n \geq kt + k + t$, the number of equations is at least the number of variables in the new linear system. As we have conjectured for virtually all cases the

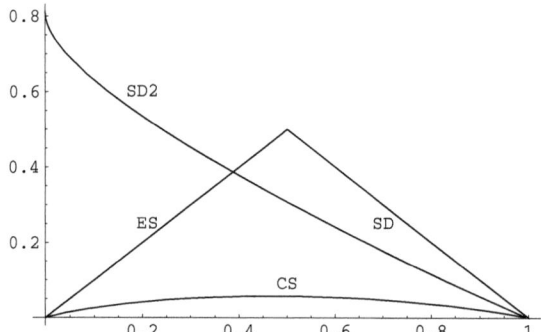

Figure 1.4.1: Exhaustive search, syndrome decoding, and covering sets algorithm. QED = our method of quadratic equations decoding, ES = exhaustive search, SD = syndrome decoding with storing coset leaders, SD2 = syndrome decoding, when t error-positions are guessed and the rest check equations are solved (as per Remark 1.3.12), CS = covering set decoding

matrix A has full rank. This means that Gaussian elimination brings A in the diagonal form that enables one to solve for the variables (that are monomials). Note that $X_1, \ldots, X_k$ are among the monomials of Sys. This implies that after applying Gaussian elimination to A in virtually all cases we get values for $X_1, \ldots, X_k$. Using then the check equations (1.12) we retrieve remaining error values $X_{k+1}, \ldots, X_n$. The complexity of Gaussian elimination is $\mathcal{O}(n^3)$ and is in fact lower if $k = \mathcal{O}(n^a), t = \mathcal{O}(n^b)$ with $a + b < 1$.

Let us continue with this idea and apply the technique of *extended linearization* ([52, 53]). We consider the binary case now, so $X_i^2 = X_i$ for all i. To illustrate the idea first consider a situation, when every equation in the system Sys is multiplied with every $X_i, 1 \leq i \leq k$. Then we obtain a new system Sys_1 that consists of $n(k+1)$ equations and these equations contain $C_1 = (t+1)\binom{k}{2} + kt + k + t + 1$ monomials. Namely, we have

1. constant monomial 1;

2. $k + t$ linear monomials $X_i, 1 \leq i \leq k$ and $Y_j, 1 \leq j \leq t$;

3. kt quadratic monomials $X_i Y_j, 1 \leq i \leq k, 1 \leq j \leq t$;

4. $\binom{k}{2}$ quadratic monomials $X_i X_j, 1 \leq i \leq k, i < j \leq k$;

5. $t\binom{k}{2}$ cubic monomials $X_i X_j Y_l, 1 \leq i \leq k, i < j \leq k, 1 \leq l \leq t$.

1.4. METHODS BASED ON QUADRATIC EQUATIONS

The terms (1) – (3) appear already in Sys, the ones from (4) – (5) are specific for Sys_1.

In the same way one can multiply the system Sys with all monomials that include $X_1, \ldots, X_k$ of degree $s \leq k$. The system, call it Sys_s obtained in this way has $n(1 + \binom{k}{1} + \cdots + \binom{k}{s}))$ equations and $C_s := C_{s-1} + \binom{k}{s+1}(t+1)$ monomials. Indeed, the system Sys_s has monomials

1. the ones from Sys_{s-1};

2. $\binom{k}{s+1}$ monomials of degree $s+1$, namely $X_{i_1} \cdot \ldots \cdot X_{i_{s+1}}$, $\{i_1, \ldots, i_{s+1}\} \subseteq \{1, \ldots, k\}, 1 \leq i_1 < \cdots < i_{s+1} \leq k$;

3. $t\binom{k}{s+1}$ monomials of degree $s+2$, namely $X_{i_1} \cdot \ldots \cdot X_{i_{s+1}} Y_j$, $\{i_1, \ldots, i_{s+1}\} \subseteq \{1, \ldots, k\}, 1 \leq i_1 < \cdots < i_{s+1} \leq k, 1 \leq j \leq t$.

Note that C_0, the number of monomials in Sys, is $(k+1)(t+1)$. Denote $\binom{k}{0} + \binom{k}{1} + \cdots + \binom{k}{s} =: f(k, s)$. So by induction $C_s = (t+1)f(k, s+1)$. Note that Sys contains equations $X_i Y_i = 0$ for $i = 1, \ldots, \min\{k, t\}$. Therefore, since we have that $X_i^2 = X_i$ for all i, this means that if we multiply $X_i Y_i$ for $i = 1, \ldots, \min\{k, t\}$ by $X_{i_1} \cdot \ldots \cdot X_{i_s}$, $\{i_1, \ldots, i_s\} \subseteq \{1, \ldots, k\}, 1 \leq i_1 < \cdots < i_s \leq k$, where one of the i_j's is equal to i, we obtain $X_{i_1} \cdot \ldots \cdot X_{i_s} Y_i$. This means that every system Sys_s repeats these equations from Sys_{s-1}. We exclude then such equations from Sys_s. In such a way we obtain that Sys_s has $nf(k, s) - \min(k, t)f(k, s-1)$ equations.

Denote the Macaulay matrix of the system Sys_s by A_s. For the further calculations we assume that A_s is full rank almost always. This is a conjectured assumption and is not proved here. Note that when one applies the XL algorithm in its full extend, cf. e.g. [130], one obtains some trivial relations that lead to linear depended rows in the extended Macaulay matrix, and therefore to non-full rank matrices. Here we only multiply the equations by X-variables taking into account the field equations over $\mathbb{F}_2$. Therefore it is not quite clear whether we will have the dependencies that exist in the full setting. Although, we have already excluded some *a priori* linear depended equations, we cannot guarantee that there are no others. Still, we believe that the degree bounds below provide a quite accurate picture for analyzing complexity for this shortened XL we consider here.

So if

$$nf(k, s) - \min(k, t)f(k, s-1) \geq C_s - 1 = (t+1)f(k, s+1) - 1,$$

then finding $X_1, \ldots, X_k$ via Gaussian elimination applied to A_s is possible. Let us give an upper bound for complexity here. Note that asymptotically in order to find $X_1, \ldots, X_k$ via extended linearization as above we need that the following holds

$$n \geq t\frac{f(k, s+1)}{f(k, s)} + \min(k, t)\frac{f(k, s-1)}{f(k, s)}. \tag{1.17}$$

We have for $N \to \infty$ and some $\lambda > 0$

$$\frac{\binom{N}{\lambda N+1}}{\binom{N}{\lambda N}} = \frac{1-\lambda}{\lambda + o(1)}.$$

We will also need that for $N \to \infty$ and $\lambda > 0$

$$\lim_{N \to \infty} \frac{1}{N} \sum_{0 \leq i \leq \lambda N} \binom{N}{i} = H(\lambda). \tag{1.18}$$

Now if $s = \lambda k$ for some $\lambda > 0$ and $k \to \infty$

$$\frac{f(k, s+1)}{f(k, s)} = 1 + \frac{\binom{k}{s+1}}{f(k, s)} = 1 + \frac{1-\lambda}{\lambda + o(1)} \frac{1}{\frac{f(k, s-1)}{\binom{k}{s}} + 1} \leq 1 + \frac{1-\lambda}{\lambda} = \frac{1}{\lambda}.$$

Moreover,

$$\frac{f(k, s-1)}{f(k, s)} < 1.$$

So if

$$n \geq \frac{t}{\lambda} + \min(k, t), \tag{1.19}$$

then the condition (1.17) is fulfilled.

Let $k = Rn, n \to \infty$ for some $0 < R < 1$, then the full error correcting capacity is $e = H^{-1}(1-R)n/2, n \to \infty$, cf. Theorem 1.1.7. Denote $\delta_e = H^{-1}(1-R)/2$, so that $e = \delta_e n$. If we consider bounded decoding, then $t \leq e$. Let $t = \delta n, \delta \leq \delta_e$. Then (1.19) yields a lower bound on λ, namely $\lambda \geq \delta/(1-\min(\delta, R))$. Thus the lowest degree (under strong assumptions of (1.19)) up to which we need to multiply with monomials including X_i is $s = \frac{\delta}{1-\min(\delta, R)} Rn$. Summing up we have the following

Proposition 1.4.71 *Taking all the above conjectures as assumptions, the upper bound on the complexity coefficient with the extended linearization as above is*

$$CC_{QED}(R) = \omega R H\left(\frac{\delta}{1-\min(\delta, R)}\right),$$

where ω is the exponent of Gaussian elimination ($2 < \omega \leq 3$).

1.4. METHODS BASED ON QUADRATIC EQUATIONS

Proof. Indeed, if we take $t = \delta n$ and $s = \delta Rn/(1 - \min(\delta, R))$, then the condition (1.19) is fulfilled and we can find $X_1 \ldots, X_k$ via Gaussian elimination applied to the matrix with $nf(k,s) - \min(k,t)f(k,s-1)$ rows and $(t+1)f(k,s+1) - 1$ columns with entries from a field extension $\mathbb{F}_Q$ with $\Theta(n)$ elements (here Q is the smallest power of 2 that is at least n). So the complexity for $\omega = 3$ is $(\log_2 n)^3 (nf(k,s) - \min(k,t)f(k,s-1))^2((t+1)f(k,s+1) - 1)$. From the point of view of the complexity coefficient only $f(k,s)^2$ and $f(k,s+1)$ matter and they are of the same order here. So using (1.18) we have

$$CC_{QED}(R) = \omega \lim_{n \to \infty} 1/n \log_2 \left(\frac{Rn}{\frac{\delta}{1-\min(\delta,R)}Rn} \right) = \omega R H_2 \left(\frac{\delta}{1 - \min(\delta, R)} \right).$$

Here we take ω rather than just 3 to take into account advanced Gaussian elimination algorithms. ◇

Experiments that we made show that actually $\delta Rn/(1 - \min(\delta, R))$ is a good estimate for the real value of s. In the following tables we investigate three cases: $R = 0.1, 0.5, 0.7$. We list n taking some values from 100 and 1000, as well as the value for error correcting capacity from Theorem 1.1.7, which is known to be quite accurate already for moderate values of n, and also the real value of s and its estimate via $\delta Rn/(1 - \min(\delta, R))$. The first table is for the case $R = 0.1$:

n	$k = \lfloor Rn \rfloor$	$e = \lfloor \delta_e n \rfloor$	s	$\delta_e Rn/(1 - \min(\delta_e, R))$
100	10	15	1	1.76
300	30	46	4	5.27
500	50	78	7	8.78
700	70	110	10	12.29
1000	100	157	14	17.56

Now the case $R = 0.5$:

n	$k = \lfloor Rn \rfloor$	$\lfloor e = \delta_e n \rfloor$	s	$\delta_e Rn/(1 - \min(\delta_e, R))$
100	50	5	2	2.91
300	150	16	8	8.73
500	250	27	13	14.55
700	350	38	18	20.38
1000	500	54	26	29.11

And finally the case $R = 0.7$:

n	$k = \lfloor Rn \rfloor$	$\lfloor e = \delta_e n \rfloor$	s	$\delta_e Rn/(1 - \min(\delta_e, R))$
100	70	2	2	1.91
300	210	7	5	5.74
500	350	12	8	9.57
700	490	18	13	13.40
1000	700	26	18	19.14

We depict the upper bound for our method as is obtained above in comparison with our benchmarks. For these figures we took $\omega = 3$. We consider three cases: $t = e, e/2, e/4$, see Figures 1.4.2, 1.4.3, 1.4.4 resp. On these figures a concatenation of exhaustive search and syndrome decoding is depicted in bold, covering sets decoding in gross dashing, and the upper estimate for our method in fine dashing.

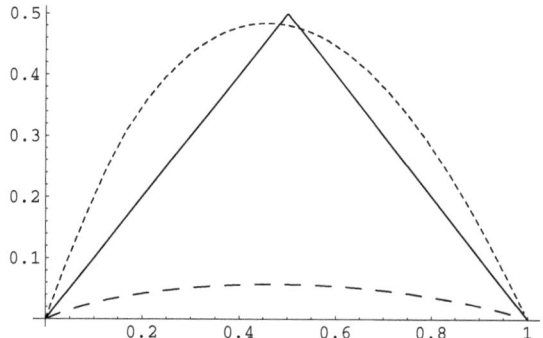

Figure 1.4.2: Exhaustive search, syndrome decoding, covering sets algorithm, and quadratic equations upper bound, $t = e$

We see that although our method performs quite poor for the case $t = e$, it becomes better for $t = e/2$ and $t = e/4$. In particular, for $t = e/4$ almost entire curves lies below the exhaustive search / syndrome decoding.

Experimental complexity comparison with random systems

Now we would like to experimentally show that our system $Sys = Sys(n, k, t)$ is solved by "general methods" in time comparable with the one needed to solve random systems of a certain sort. Let us be more specific. As has already been noted the system $Sys(n, k, t)$ has the following monomials

1.4. METHODS BASED ON QUADRATIC EQUATIONS

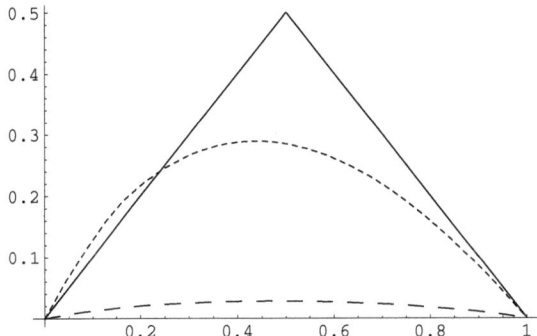

Figure 1.4.3: Exhaustive search, syndrome decoding, covering sets algorithm, and quadratic equations upper bound, $t = e/2$

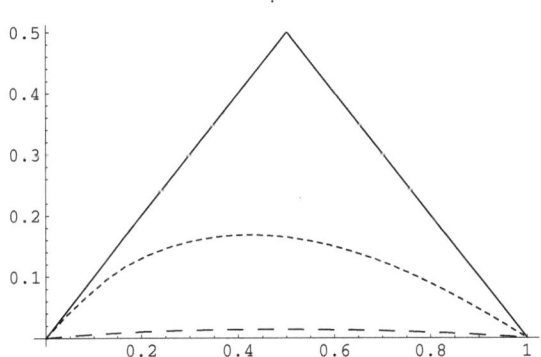

Figure 1.4.4: Exhaustive search, syndrome decoding, covering sets algorithm, and quadratic equations upper bound, $t = e/4$

1. constant monomial 1;
2. $k + t$ linear monomials $X_i, 1 \leq i \leq k$ and $Y_j, 1 \leq j \leq t$;
3. kt quadratic monomials $X_i Y_j, 1 \leq i \leq k, 1 \leq j \leq t$.

Denote by R_1 a system of n quadratic equations that has the monomials listed above, but the corresponding coefficients are randomly taken from $\mathbb{F}_Q$.

We also require that R_1 has a unique solution in the algebraic closure of $\mathbb{F}_Q$. Denote by R_2 a system that has the same properties as R_1, but the requirement on uniqueness of a solution is dropped out. Note that such systems do not have solutions in general. Finally, denote by R_3 a fully random system of n quadratic equations, i.e. it has all possible monomials of degree ≤ 2 and the corresponding coefficients are randomly taken from $\mathbb{F}_Q$. Such systems also do not have solutions in general.

The following table shows some examples of solving the above systems in SINGULAR for different parameters (n, k, t). The time is given is seconds.

(30, 10, 6)				(40, 10, 8)				(65, 15, 8)			
Sys	R_1	R_2	R_3	Sys	R_1	R_2	R_3	Sys	R_1	R_2	R_3
5	5	5	220	12	12	13	572	33	39	38	5076

So, as we see the timings are so that the following conjecture may be stated.

Conjecture 1.4.72 *We have the following relations between the complexities for solving Sys, R_1, R_2, and R_3 with "general methods"*

$$Compl(Sys) \approx Compl(R_1) \approx Compl(R_2) << Compl(R_3).$$

By this Conjecture we may try to estimate the complexity of solving Sys by the estimates available for systems R_3. The systems R_3 have a lot to do with the semi-regular sequences [10, 11, 12]. It turns out that estimates based on R_3-systems are too gross. On the other hand, possible estimates available for the R_2- or R_1-systems should yield pretty good estimates for $Compl(Sys)$.

1.4. METHODS BASED ON QUADRATIC EQUATIONS

Degree of polynomials during GB computations

Now we discuss how high can be the degree of the polynomials that appear during GB computations for our systems. We briefly discuss here results for SINGULAR implementation of Buchberger's algorithm and its modifications, MAGMA's F4 implementation and theoretical bound.

In the table below we compare highest degrees when computing GB with respect to degree reverse lexicographic order `degrevlex` with `std` command in SINGULAR. For every code the left column shows degree for the system as per Corollary 1.4.55, and the right one for the system as per Definition 1.4.37.

no. of err.	[120,40]		[120,30]		[120,20]		[120,10]		[150,10]	
2	4	4	3	3	3	3	3	3	3	3
3	7	6	5	5	3	3	3	3	3	3
4	6	5	5	6	3	3	3	3	3	3
5	-	-	6	5	6	6	3	3	3	3
6	-	-	5	5	6	5	3	3	3	3
7	-	-	5	5	5	5	3	3	3	3
8	-	-	-	-	5	5	3	3	3	3
9	-	-	-	-	5	5	3	3	3	3
10	-	-	-	-	5	5	3	3	3	3
11	-	-	-	-	33	10	6	6	3	3
12	-	-	-	-	-	-	6	5	3	3
13	-	-	-	-	-	-	5	5	4	4
14	-	-	-	-	-	-	5	5	5	6
15	-	-	-	-	-	-	5	5	6	5
16	-	-	-	-	-	-	5	5	5	5
17	-	-	-	-	-	-	5	5	6	5
18	-	-	-	-	-	-	5	5	5	5
19	-	-	-	-	-	-	5	5	5	5
20	-	-	-	-	-	-	6	6	5	5
21	-	-	-	-	-	-	6	5	5	5
22	-	-	-	-	-	-	-	-	5	5
23	-	-	-	-	-	-	-	-	5	5
24	-	-	-	-	-	-	-	-	5	5

Here "-" means that computation takes more than 1000 sec. Other cells correspond to what we call here "feasible" computations.

Note that for both systems degrees do not quite go in an ascending manner. The case $(120, 20, 11)$ is also quite interesting as here we observe a hike in degree. Even more notable is the fact that with `slimgb` command from SINGULAR we never obtained degrees higher than 3 (compare to the above), and still the running times were really slower. Therefore, we may conclude that the highest degree alone does not determine time complexity. For MAGMA F4 implementation (`GroebnerBasis` command) we observed more stable behavior and the highest degrees increasing from 3 to 4 for the feasible cases. By computing the real value of s we obtain that highest degrees of the extended linearization for the feasible computations do not go higher than 3 except for the case $(120, 20, 11)$, where it is 4. For many feasible cases direct linearization was feasible.

1.4.8 Simulations and experimental results

All computations in this subsection are undertaken on AMD Opteron Processor 242 (1.6MHz), 8GB RAM under Linux. The computations of Gröbner bases are realized in SINGULAR 3-0-3 [74]. The command used is `std`. For Gröbner basis computations we used degree reverse lexicographic order (`dp`). For the generation of equation systems and processing the SINGULAR-library `decodegb.lib` [30] is used, see for the user's manual [123]. For the construction we take the field extension $\mathbb{F}_q$ with the smallest q not smaller than n. Next, as an MDS matrix we use Vandermonde matrix $B(\mathbf{a})$, as per Definition 1.4.9, where $\mathbf{a} = (a, a^2, \ldots, a^n)$ for some primitive element a.

Random binary codes

Experiment 1.4.73 Here we present some results on decoding with the use of Theorem 1.4.45 for binary random codes. First we determine the minimum distance of a random code with Theorem 1.4.61 and then perform decoding of some given number of received words. The number of errors that occur in these received words equals the error capacity of the code. The results are given in the following table. In the columns are the parameters of the code, the error-correcting capacity, time to compute the minimum distance, total time to decode with Gröbner bases, the number of received words, and the average time to decode one word with Gröbner bases, respectively. The time is given in seconds.

1.4. METHODS BASED ON QUADRATIC EQUATIONS 79

Code	err. cap.	mindist.	GB dec.	no. of rec.	average
[25,11,4]	1	2.99	1.10	300	0.0037
[25,11,5]	2	21.58	2.89	300	0.0096
[25,8,5]	2	0.99	1.84	300	0.0061
[25,8,6]	2	3.38	1.79	300	0.0060
[25,8,7]	3	12.26	6.94	300	0.0231
[31,15]	2	-	10.76	300	0.0359
[31,15]	3	-	11.19	10	1.119

The bar "-" means that a computation took more than 1000 sec. so we were not able to actually compute the minimum distance in a short time, so we have just assumed the error capacity.

Experiment 1.4.74 We are able to correct even more errors in larger codes. The following table shows timings for random binary [120, 10], [120, 20], [120, 30] and [150, 10] codes, where 1 means one second or less. We also present here timings for Magma's F4 [38] in order to show that our approach actually does not depend on the concrete Gröbner basis algorithm. Decoding time appears to have low variance among error-vectors of the given weight. In the table below, for every code the left column corresponds to SINGULAR and the right column to Magma.

no. of err.	[120,40]		[120,30]		[120,20]		[120,10]		[150,10]	
2	1	1	1	1	1	1	1	1	1	1
3	22	7	1	1	1	1	1	1	1	1
4	172	64	5	14	1	1	1	1	1	1
5	804	228	31	36	1	1	1	1	1	1
6	-	-	98	63	3	9	1	1	2	1
7	-	-	471	144	7	15	1	1	2	1
8	-	-	-	-	17	25	1	1	2	1
9	-	-	-	-	43	38	1	1	2	1
10	-	-	-	-	109	51	1	1	2	1
11	-	-	-	-	392	84	1	1	3	1
12	-	-	-	-	-	630	2	8	3	1
13	-	-	-	-	-	-	2	9	4	1
14	-	-	-	-	-	-	3	11	4	1
15	-	-	-	-	-	-	7	13	5	20
16	-	-	-	-	-	-	10	16	5	22
17	-	-	-	-	-	-	22	19	8	26

CHAPTER 1. SYSTEM SOLVING IN DECODING

continued from previous page

no. of err.	[120,40]		[120,30]		[120,20]		[120,10]		[150,10]	
18	-	-	-	-	-	-	38	23	8	30
19	-	-	-	-	-	-	72	28	16	38
20	-	-	-	-	-	-	183	33	27	43
21	-	-	-	-	-	-	265	48	43	50
22	-	-	-	-	-	-	362	64	69	59
23	-	-	-	-	-	-	688	723	128	69
24	-	-	-	-	-	-	-	-	261	82
25	-	-	-	-	-	-	-	-	575	93

Experiment 1.4.75 Now we consider the situation of the "true" decoding, i.e. when we do not know the number of errors that occurred. So then we have to solve all the systems $J(i, \mathbf{r})$ for $i = 1, \ldots, e$, where e is the actual number of errors, as per Algorithm 1.4.49. In the following table we show cumulative time needed to solve the first $e-1$ systems, then separately time needed for solving the system $J(e, \mathbf{r})$ and the ratio between the two. It is remarkable that it makes no difference in time on whether the system $J(i, \mathbf{r})$ has a unique solution or has no solution for given i. The notation below is $[n, k, e]$.

[120,40,4]	[120,30,7]	[120,20,11]	[120,10,21]	[150,10,24]
14	273	312	628	651
313	933	633	483	513
0.04	0.29	0.49	1.3	1.27

It is seen from the table that the ratio increases as the information rate of a code decreases.

We do the same for the minimum distance finding. In the following table we show cumulative time needed to solve the first $d-1$ systems (d is the minimum distance), then separately time needed for solving the system $J(d, 0)$ and the ratio between the two.

[25,15,3]	[30,10,7]	[30,15,5]	[40,15,7]
1	11	69	422
15	46	982	2478
0.07	0.24	0.07	0.17

Here we see that time needed for solving $J(d, 0)$ strongly dominates all the previous work needed.

1.4. METHODS BASED ON QUADRATIC EQUATIONS

Experiment 1.4.76 In order to speed up the computation we applied the following trick. We perform Gaussian elimination for the system in $J(\mathbf{r})$ and express some $n - k$ syndrome variables via k others. We then substitute expressions for these $n - k$ variables in the system from $I(t, \mathcal{U}, V)$ and perform Gaussian elimination with variables (w.l.o.g.) $U_1, \ldots, U_k$, $V_1, \ldots, V_t$, and $U_i V_j, 1 \leq i \leq k, 1 \leq j \leq t$. We add the resulting syzygies to the original system (i.e. $J(t, \mathbf{r})$) and then compute the reduced Gröbner basis. We order the variables in such a manner that these syzygies have as many linear terms as possible. Corresponding timings follow.
For the codes $[120, 20]$:

no. of err.	6	7	8	9	10	11
preprocessing	2	3	4	4	4	4
solve	3	7	16	33	65	550
total	5	10	20	37	69	554
before	5	14	32	74	183	633

Here "preprocessing" means two Gaussian eliminations and we excluded time needed for substitution of expressions.
For the codes $[120, 10]$:

no. of err.	14	15	16	17	18	19	20	21
preprocessing	3	3	3	4	4	4	4	4
solve	5	6	8	9	12	15	59	295
total	8	9	11	14	16	19	63	299
before	6	14	20	29	71	139	327	483

For the codes $[150, 10]$:

no. of err.	15	16	17	18	19	20	21	22	23
preprocessing	5	6	6	7	7	9	9	12	14
solve	9	11	13	15	16	18	22	27	37
total	14	17	19	22	23	27	31	39	52
before	10	11	16	16	34	53	84	153	241

Experiment 1.4.77 In this experiment we consider how complexity of solving bounded decoding systems depends on the information rate of a code, provided that the fraction of errors we want to correct is fixed. We consider two cases of random binary codes: $n = 120$ and $n = 100$. In both cases the fraction of errors to correct is set to be $1/2$ of the error-correcting capacity

(corresponding numbers are obtained via Theorem 1.1.7). The results are depicted on the Figure 1.4.5. The y-axis corresponds to the time in seconds, and x-axis to the information rate.

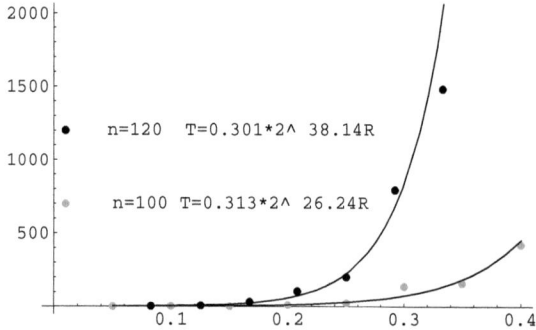

Figure 1.4.5: Decoding binary random codes with the fixed fraction of errors to correct

Along with the experimental figures we provide also the best fit for the data. Namely we were looking for the best fit among the functions of the form $f(x) = a \cdot 2^{bx}$. It seems that these functions fit the data pretty well.

So we see that the complexity increases exponentially, when the redundancy of a code decreases. Moreover the complexity gets worse when we go to larger codes.

Experiment 1.4.78 Let us compare our method with the method of Fitzgerald-Lax [70]. We follow the same pattern and notation of codes as in Experiment 1.4.73. First, let us take a look at "small" codes.

code	err. cap.	GB dec.	no. of rec.	average
[25,11]	1	0.32	300	0.0011
[25,11]	2	14.48	300	0.0483
[25,8]	2	6.03	300	0.0201
[25,8]	3	4.68	1	4.68
[31,15]	2	11.46	100	0.1146
[31,15]	3	112.14	1	112.14

So, we see that except for the case of [25, 11] code with 1 error, our method wins, sometimes substantially (cf. [25, 8], [31, 15] with 2 errors, and in particular [31, 15] with 3 errors).

1.4. METHODS BASED ON QUADRATIC EQUATIONS 83

The difference is even more striking when working with $[120, 10], [120, 20], [120, 30]$ and $[150, 10]$ codes.

no. of err.	[120,30]	[120,20]	[120,10]	[150,10]
2	5	2	1	2
3	3996	2263	1544	804

These simulations indicate that when dealing with random (binary) codes this method [70], has problems starting already at 3 errors.

Hermitian codes

The situation is a bit different if we try to compare both methods on, e.g. Hermitian codes. In the case of the Hermitian curve $y^2 + y = x^3$ over $\mathbb{F}_4$ both methods correct fast up to correction capacity. Interesting things start to happen, when working with the Hermitian curve $y^3 + y = x^4$ over $\mathbb{F}_9$. If we consider a $[27, 18, \geq 7]$ code which corrects 3 errors, then FL method corrects them in less than a second, whereas our method needs 1586 sec. (938 sec. when using Gaussian elimination trick as above). This can be explained with the fact that FL method is better adjusted here to the Hermitian case by including the curve equation in (1.10), whereas our method is a generic one and loses this information. Nevertheless, for the $[27, 8, \geq 17]$ code we obtain the following:

no. of err.	3	4	5	6
our method	0	1	5	43
FL method	0	1212	-	-

So, here our method corrects pretty fast up to 6 errors, whereas FL method gets slow already for 4 errors.

Cyclic codes

Now we take a look at cyclic codes.

Experiment 1.4.79 The defining set of a cyclic code is given in the second column. Full error-correcting capacity was used.

code	roots	dec.	no. of rec.	average

CHAPTER 1. SYSTEM SOLVING IN DECODING

continued from previous page

code	def.set	dec.	no. of rec.	average
[15,7,5]	1,3	0.555	1000	0.0005
[15,5,7]	1,3,5	0.885	1000	0.0008
[31,16,7]	1,5,7	6.05	1000	0.0060
[33,12,10]	0,1,5	63	100	0.63
[35,17,6]	1,5,15	4.7	100	0.047
[35,16,7]	1,5,7	84	100	0.84
[39,24,6]	0,1,13	179	100	1.79
[39,15,10]	1,3	147	100	1.47
[41,21,9]	1	491	1	491
[41,20,10]	0,1	182	1	182
[45,23,7]	1,3,5	180	10	18.0

First of all, we note a hike in computational time, e.g. between [35,17,6] and [35,16,7] codes, which appear to differ only slightly. Note also the hike between similar codes [39,15,10] and [41,21,9]. Here the fact that we use Vandermonde matrix as an MDS matrix plays a crucial role, see Experiment 1.4.81.

Experiment 1.4.80 Now we compare our method with the method by Augot *et.al.* based on Waring function (cf. Section 1.3.2). We use SINGULAR's command `slimgb` as more preferable here. The corresponding table is as follows:

code	def.set	dec.	no. of rec. words	average
[15,7,5]	1,3	0.253	1000	0.0002
[15,5,7]	1,3,5	0.462	1000	0.0004
[31,16,7]	1,5,7	2.96	1000	0.0029
[35,17,6]	1,5,15	0.25	100	0.0025
[35,16,7]	1,5,7	0.65	100	0.0065
[39,15,10]	1,3	1.73	100	0.0173
[41,21,9]	1	34	100	0.34
[45,23,7]	1,3,5	0.94	100	0.0094

We see an overwhelming superiority of this method over ours (Experiment 1.4.79), except of the case with the codes [15,5,7], [15,7,5], and [31,16,7], where both methods are of the same magnitude.

1.4. METHODS BASED ON QUADRATIC EQUATIONS 85

Experiment 1.4.81 Next we would like to consider a situation, when the MDS matrix B is chosen not with the rule "2^m is the smallest, such that $2^m > n$", but with the rule usual for cyclic codes "2^m is the smallest, such that $n | (2^m - 1)$". In this situation B becomes an RS-matrix. As a result $I(t, \mathcal{U}, V)$ has a form of the GNI, cf. Section 1.4.4. The results follow

code	def.set	dec.	no. of rec. words	average
[35,17,6]	1,5,15	3.3	100	0.033
[35,16,7]	1,5,7	63	100	0.63
[39,24,6]	0,1,13	152	100	1.52
[39,15,10]	1,3	103	100	1.03
[41,21,9]	1	204	100	2.04
[41,20,10]	0,1	172	100	1.72
[45,23,7]	1,3,5	0.69	100	0.0069

The codes $[15, 7, 5], [15, 5, 7], [31, 16, 7]$ are not included here as the coefficient field remains the same. Next we observe that for the cases $[45, 23, 7]$, $[41, 20, 10], [41, 21, 9]$ a quite tremendous speed-up is obtained, whereas for the rest of the cases we do not observe much of an improvement. Note that for all cases except for $n = 41$, the field extension was $\mathbb{F}_{12}$. For this field SINGULAR has look-up table arithmetics. For the case $n = 41$ the field extension is $\mathbb{F}_{20}$ and SINGULAR does not have a look-up anymore. So considering straightforward possible optimizations, we may state that the cases $[45, 23, 7], [41, 21, 9]$ are now comparable with the Waring function analogues.

Beyond error capacity

Up to now we only considered the situation of bounded decoding, i.e. correcting errors up to the error-correcting capacity of a code. Now we want to move further and use Theorem 1.4.42 to its full potential. Namely we will find closest codewords to a received word, when the number of errors occurred is larger than error capacity. In the following table we list results for the cyclic codes we considered above. We take care that if the number of errors occurred is $t > e$, then there is no codeword at distance less than t to a given received word. We used 5 samples for each code to derive the average value. The values in each sample did not differ significantly. We write "-" either when $t \leq e$ or if the computation took more than 500 secs.

code	err.cap.	\multicolumn{5}{c}{t, number of errors occurred}				
		3	4	5	6	7
[15,7,5]	2	0.0013	-	-	-	-
[15,5,7]	3	-	0.0011	0.0023	-	-
[31,16,7]	3	-	0.353	83.6	-	-
[33,12,10]	4	-	-	1.82	17.5	154.9
[35,17,6]	2	1.78	70.0	-	-	-
[35,16,7]	3	-	15.9	323.7	-	-
[39,24,6]	2	52.0	-	-	-	-
[39,15,10]	4	-	-	41.4	-	-

Remarks

Remark 1.4.82 We note that the rate of a code is a determining factor for complexity. Indeed, we have a system with $n + t$ variables and $n + r$ equations. It was noticed by researchers that overdetermined systems of algebraic equations in general are easier to solve (cf. e.g. [10], [11], [120]). So if, for given n, we increase redundancy r, or reduce the number of errors t we want to solve, the system becomes more overdetermined, which positively reflects on complexity. We could see on the above tables, how decrease in dimension caused better performance of the system.

Experiment 1.4.83 Let us now make some remarks on the "classical" syndrome decoding. One version of syndrome decoding is implemented for example in the GAP computer algebra system [71]. There coset leaders (c.l.) are explicitly computed and stored in a table for the further decoding, see [82], sections 4.10-1 and 4.10-9. So, the major part of time is spent during the first decoding (when the table is precomputed), whereas further it takes almost no time. Also here the method is independent on t. We have the following (for binary random codes):

code	[25,11]	[25,8]	[31,15]
time for c.l. computation, sec	1.8	15.5	8.0

Already for a random binary code [35, 15] GAP is not able to perform decoding and returns an error. Similar performance was shown by MAGMA computer algebra system [38]. We were unable to handle syndrome decoding over $\mathbb{F}_2$, when a redundancy $n - k$ exceeded 20. So as we see, the syndrome decoding can be effective only in case of small values of $n - k$, whereas our method provides a better flexibility with respect to these parameters.

Conclusions and future work

In this chapter we presented a method for decoding linear codes based on a system of quadratic equations. We have seen that it is possible to solve the decoding problem with the quadratic system without adding field equations both for bounded distance decoding and nearest codeword decoding. We took a look at the adaptations of the method to the finding of the minimum distance and complexity issues, as well as computer simulations.

The following we consider as promising directions in this area.

- In Section 1.4.4 we showed how one can obtain generalized Newton identities for arbitrary linear codes similar to the well-known identities existing for the cyclic codes (Section 1.3.2). Note that cyclic codes are given via parity check matrices that are submatrices of the form (1.1). This yields a lot of nice structure, in particular one can compute certain unknown syndromes at no cost, cf. Proposition 1.4.57. Similarly, if one considers codes that are given via parity check matrices that are submatrices of some MDS matrices, this can simplify our method of quadratic equations. Namely, knowledge of certain unknown syndromes may significantly simplify the quadratic part: recall that under certain assumptions in the cyclic case one can even obtain a linear system with a unique solution. Studying such families of codes and effective decoding algorithms seems a very promising direction here. Note that since a choice of an underlying MDS matrix is vast, one can obtain many codes of such sort.

- Complexity issues touched on in Section 1.4.7 should definitely be studied further. Namely, it is important to close the statements that are stated as conjectures there. More advanced techniques than the weak extended linearization should also be considered: they may provide better estimates of complexity.

- Equivalent representation of our system given in Theorem 1.4.54 and Corollary 1.4.55 provides better understanding of the method and gives an opportunity to employ more coding-theoretic methods in the considerations. Properties and corollaries from such constructions should be studied further.

- The question of radicality of $J(t, \mathbf{r})$ is interesting from the point of view of computational algebra.

Chapter 2
System solving in cryptanalysis

2.1 Introduction

In this chapter we investigate several methods for obtaining and solving key variables only equations, which appear in cryptanalysing the small scale variants of the Advanced Encryption Standard (AES). The cipher Rijndael was chosen as the AES in 2001, and was published as a FIPS 197 standard [102]. AES provides fast and simple symmetric encryptions, while maintaining high resistance to (known) attacks. Simplicity of the AES was criticized since the moment it had appeared, but no one has proposed an attack, which would, at least theoretically, break the AES. We will concentrate on the so-called algebraic attacks. They emerged in the papers of Courtois [54] and Murphy et.al. [101, 46]. The idea is to present the action of the AES block cipher as a system of algebraic equations over a finite field. Then, solving such a system would reveal a secret key. In [54] Courtois and Pieprzyk construct a system over $\mathbb{F}_2$, whereas in [101] Murphy et. al. propose to consider the system over $\mathbb{F}_{2^8}$. Courtois obtained his system directly from the AES, wheres Murphy et. al. proposed an embedding of the AES state space to a larger space. Different manipulations and variations of these methods were considered since their appearance. Some works in this area are [1, 2, 44, 45, 46, 47, 28, 29, 54, 43, 101, 109, 127]. So far no method presented any real threat to AES. For better understanding and in order to facilitate experimenting, small scale variants of AES were proposed [47].

We elaborate on the initial proposals. The fact that the equations mentioned above have many auxiliary variables, complicates the use of several plaintext/ciphertext pairs. So we aim at obtaining systems that contain only variables responsible for an unknown key. We use the fact that most part of the equations above already constitutes a Gröbner basis for a suitable ordering and then apply normal form computations. The idea of using the zero-dimensional Gröbner-representation for AES was first proposed in [28] and also considered in [127]. The main difference of our approach from the former is that we work over $\mathbb{F}_2$ and we include the field equations still being able to exploit the Gröbner structure; we obtain equations in key-variables only and solve them. The latter approach is more similar to ours, but the difference is that again we work with equations over $\mathbb{F}_2$, rather than with the BES equations as in [127]. In this respect we believe that optimized data structures of POLYBORI[24], which is the main tool in our experiments, yield an advantage. Also we provide a solid experimental material, which was not done in [127].

The chapter is organized as follows. In Section 2.2 we provide an overview of algebraic cryptanalysis of block ciphers. Then in Section 2.3 we give a brief description of the Advanced Encryption Standard. The core of the chapter is Section 2.4. First in Section 2.4.1 we analyze the structure of equations that come from algebraic cryptanalysis of AES and rewrite the system for our method. In Section 2.4.2 we exploit the Gröbner structure of our new system and do some experiments. Next we turn out attention to the meet-in-the-middle attack in Section 2.4.3. Some refinements of our method in the meet-in-the-middle scenario are given in Section 2.4.4. Necessary notions from computational algebra are in Section 1.2.

2.2 Block ciphers and algebraic cryptanalysis

In this chapter we will be dealing with the block cipher Rijndael [56] which was adopted for the Advanced Encryption Standard (AES) on 2001 [102]. The AES is used mostly for symmetric cryptographic communication. In this section we give a very brief introduction to symmetric cryptosystems, block ciphers, and algebraic cryptanalysis. More on symmetric cryptography in [27, 98, 125].

The idea behind symmetric cryptography is quite simple and thus was basically known for quite a long time. The task is to convey a secret between two parties, called traditionally Alice and Bob, so that figuring the secret out is not possible without knowledge of some additional piece of information. This additional piece of information is called a secret key and is supposed to be known only to the two communicating parties. The secrecy of the transmitted message rests entirely upon the knowledge of this secret key, and thus if an adversary or an eavesdropper, traditionally called Eve, is able to find out the key, then the whole secret communication is corrupted. Now let us take a look at the formal definition.

2.2. BLOCK CIPHERS AND ALGEBRAIC CRYPTANALYSIS

Definition 2.2.1 The *symmetric cryptosystem* is defined by the following data:

- The *plaintext space* $\mathcal{P}$ and the *ciphertext space* $\mathcal{C}$.

- $\{E_e : \mathcal{P} \to \mathcal{C} | e \in \mathcal{K}\}$ and $\{D_d : \mathcal{C} \to \mathcal{P} | d \in \mathcal{K}\}$ are the sets of *encryption* and *decryption* transformations, which are bijections from $\mathcal{P}$ to $\mathcal{C}$ and from $\mathcal{C}$ to $\mathcal{P}$ resp.

- The above transformations are parametrized by the *key space* $\mathcal{K}$.

- Given an associated pair (e, d), so that a property $\forall p \in \mathcal{P} : D_d(E_e(p)) = p$ holds, knowing e it is "computationally easy" to find out d and *vise versa*.

The pair (e, d) is called the *secret key*. Moreover, e is called the *encryption key* and d is called the *decryption key*.

Note that often the counterparts e and d coincide. This gives a reason for the name "symmetric". Let us take a look at a concrete example of a symmetric cryptosystem.

Example 2.2.2 The first use of a symmetric cryptosystem is conventionally attributed to Julius Caesar. He used the following cryptosystem for communication with his generals, which is historically called *Caesar cipher*. Let $\mathcal{P}$ and $\mathcal{C}$ be the sets of all strings composed of letters from the English (Latin for Caesar) alphabet $\mathcal{A} = \{A, B, C, \ldots, Z\}$. Let $\mathcal{K} = \{0, 1, 2, \ldots, 25\}$. Now an encryption transformation E_e given a plaintext $p = (p_1, \ldots, p_n), p_i \in \mathcal{A}, i = 1, \ldots, n$ does the following. For each $i = 1, \ldots, n$ one determines a position of p_i in the alphabet $\mathcal{A}$ ("A" being 0, "B" being 1, ..., "Z" being 25). Next one finds a letter in $\mathcal{A}$ that stands e positions to the left, thus finding a letter c_i; one needs to overlap if the beginning of $\mathcal{A}$ is reached. So with the enumeration of $\mathcal{A}$ as above, we have $c_i = p_i - e$ (mod 26). In this way a ciphertext $c = (c_1, \ldots, c_n)$ is obtained. Decryption key is given by $d = -e$ (mod 26), or, equivalently, for decryption one needs to shift letters e positions to the right.

Julius Caesar used $e = 3$ for his cryptosystem. Let us consider an example. For the plaintext $p =$"BRUTUS IS AN ASSASSIN", the ciphertext (if we ignore spaces during the encryption) looks like $c =$"YORQRP FP XK XOOXOOFK". To decrypt one simply shifts back 3 positions to the right.

The above is a simple example of a so-called *substitution cipher*, which is in turn an instance of a *block cipher*. Block ciphers, among other things, provide a practical realization of symmetric cryptosystems. They can also be used for asymmetric cryptosystems and constructing other cryptographic primitives, like pseudorandom number generators, authentication codes, hash functions. The formal definition follows.

Definition 2.2.3 The n-bit *block cipher* is defined as the mapping $E : \{0,1\}^n \times \mathcal{K} \to \{0,1\}^n$, where $\mathcal{K}$ is the *key space* (usually $\mathcal{K} \subseteq \{0,1\}^l$ for some l) and for each $k \in \mathcal{K}$ the mapping $E(\cdot, k) =: E_k : \{0,1\}^n \to \{0,1\}^n$ is invertible. E_k is the *encryption transformation* for the key k, and $E_k^{-1} = D_k$ is the *decryption* transformation. If $E_k(p) = c$, then c is the *ciphertext* of the *plaintext* p under the key k.

So like in the case of block codes, one first has to partition a message to be sent into blocks of length n, and then apply a block cipher transformation blockwise. Other possibilities exist out there: one can process information as a *stream*. This idea is realized by *convolutional codes* in coding theory and by *stream ciphers* in cryptography.

Example 2.2.4 (*Permutation or transposition cipher*) The idea of this cipher is to partition the plaintext into blocks and perform a permutation of elements in every block. More formally, partition the plaintext into blocks of the form $p = p_1 \ldots p_n$ and then permute: $c = E_k(p) = p_{k(1)}, \ldots, p_{k(n)}$. The key space $\mathcal{K}$ now is the set of all permutations of $\{1, \ldots, n\} : \mathcal{K} = S_n$. For example let the plaintext be $p =$"CODING AND CRYPTO", let $t = 5$, and $k = (4,2,5,3,1)$. If we remove the spaces and partition p into 3 blocks we obtain $c =$"IONDCDACNGTYOPR".

The block ciphers in Examples 2.2.2 and 2.2.4 are easy to break. Nevertheless, a clever combination of the two ideas above is used in modern ciphers to provide high security level. Namely, the permutation parts used to provide *diffusion* and substitution parts to provide *confusion* in a ciphertext. Confusion is usually realized through a layer of *S-Boxes*, carefully chosen non-linear transformations that act on parts of the state vector from $\{0,1\}^n$.

Most of the modern block ciphers are iterative ciphers.

Definition 2.2.5 An *iterative* block cipher is a block cipher which performs sequentially a certain transformation F. This transformation is called *round transformation* and the number of rounds N_r is a parameter of an iterative cipher. It is also common to *expand* the initial private key k to subkeys $k_i, i = 1, \ldots, N_r$, where each k_i is used as a key for F at round i. A procedure for obtaining the subkeys from the initial key is called a *key schedule*. For each k_i the transformation F should be invertible to allow decryption.

2.2. BLOCK CIPHERS AND ALGEBRAIC CRYPTANALYSIS

The idea of confusion/diffusion lies behind a bulk of block ciphers including Data Encryption Standard (DES) that was used since 1976 until mid. 1990s and AES. DES operates on 64-bit message blocks, whereas key size is 56 bit. The successor of DES, AES operates on 128-bit blocks, the key size may be 128, 192, or 256 bits. We give a brief description of AES in the next section. Other well-known block ciphers are Serpent, Blowfish, and Twofish, which lost the AES competition to Rijndael. Operation block sizes of these ciphers are the same as for the Rijndael-AES.

Since we are going to be attacking certain instances of AES it is important to know which assumptions on eavesdropper's power are made.

Assumptions:

- The eavesdropper has an access to all ciphertexts that are transmitted through the communication channel. He/she is able to extract these ciphertexts and use them further for his/her disposal.

- The eavesdropper has a full description of the block cipher itself, i.e. he/she is aware of how the encryptions constituting the cipher act.

Also we have to specify possible scenarios of crypto attacks.

Attack scenarios:

- *ciphertext-only:* The eavesdropper does not have any additional information, only an intercepted ciphertext.

- *known-plaintext:* Some amount of plaintext-ciphertext pairs are available to the eavesdropper.

- *chosen-plaintext and chosen-ciphertext:* The eavesdropper has an access to plaintext-ciphertext pairs with a specific eavesdropper's choice of plaintexts and ciphertexts resp.

- *adaptive chosen-plaintext and adaptive chosen-ciphertext:* The choice of the special plaintexts resp. ciphertext in the previous scenario depends on some prior processing of pairs.

- *related-key:* The eavesdropper is able to do encryptions with unknown yet related keys, with the relations known to the eavesdropper.

In this thesis we are dealing with the known-plaintext scenario.

Below we sketch some basic ideas of algebraic cryptanalysis of block ciphers. A good recent survey on this topic can be found in [42]. Since a block cipher maps plaintexts to ciphertexts (which are usually both binary vectors), one can apply interpolation techniques to describe work of a cipher via a polynomial of sufficiently high degree with unknown coefficients. Having then necessary amount of plaintext/ciphertext pairs, one, in theory, is able to compute these coefficients. These were initial observations by researchers. In practice it turned out that such a representation is highly inefficient, since the describing polynomial is out of reach for computing. After realizing that researchers switched to considering *systems of equations*, rather than one equation. The idea is not new, however. In fact it stems from the pioneering work of Claude Shannon [121] where he proclaims that breaking a cryptosystem should require "as much work as solving a system of simultaneous equations in a large number of variables". It should not come as a surprise that one can describe a particular block cipher as an equation system in many ways. This gives a researcher a freedom of choice and an opportunity to exploit certain peculiarity of a system he/she has constructed. The systems constructed may vary in the number of variables, highest degree, coefficient field, in extent to which they are sparse etc. For example, since block ciphers usually work with binary vectors, it is possible to construct a system over the field $\mathbb{F}_2$. It is also may be possible, though, to construct it over the field $\mathbb{F}_{2^m}$, where m is the size of vectors an S-Box operates on. In this way, one can gain in sparsity of the system. In the case of AES in [54] the authors present a quadratic system over $\mathbb{F}_2$ with 8000 equations and 1600 variables, whereas in [101] a quadratic system over $\mathbb{F}_{2^8}$ is presented with 7808 equations and 4608 variables. The latter system is much sparser that its $\mathbb{F}_2$-counterpart.

Following [42] we distinguish between two types of systems for algebraic cryptanalysis of block ciphers. The first one is *implicit*. Here the equations are of the form

$$g_l(X_{l+1}, X_l, K_l) = 0.$$

Thus the state variables of the next round are implicitly described via the previous state variables and key variables for that round. The second one is *explicit*. Here the equations are of the form

$$X_{l+1} = f_l(X_l, K_l).$$

Most of the cryptanalytic research has been done on the systems of the first type. In this thesis the address the second type.

2.3 AES: Advanced Encryption Standard

In this section we give a brief overview of the Advanced Encryption Standard (AES). For the full description of the AES we refer to [102]. AES in its standard form (the so-called AES-128) operates on rectangular arrays of bytes.

2.3. AES: ADVANCED ENCRYPTION STANDARD

So all operations are performed on the 4 × 4 arrays of bytes. As we have already mentioned, the AES is composed of relatively simple operations in order to ensure its efficient implementation. A set of initial operations that are being executed consecutively composes a *round*. AES-128 performs 10 rounds, where 9 rounds are the same, and the last 10*th* round differs a little. A byte is considered either as an element of $\mathbb{F}_{2^8}$ or as a $\mathbb{F}_2$-vector of length 8 via $\mathbb{F}_{2^8} = \mathbb{F}_2[a]/\langle m(a)\rangle$, where $m(a) = a^8 + a^4 + a^3 + a + 1$ is the *Rijndael polynomial*. The specifications of such a transformation are given in [102]. The following algebraic description of one round of the AES is from [101].

The AES S-Box. The value of each byte in the array is substituted according to a table look-up. A result of this table look-up $S[\cdot]$ is the combination of three transformations.

- The input w considered as an element from $\mathbb{F}_{2^8}$ and is mapped to $x = w^{(-1)}$, where $w^{(-1)}$ is defined by

$$w^{(-1)} = w^{254} = \begin{cases} w^{-1} & w \neq 0, \\ 0 & w = 0. \end{cases}$$

Thus "AES inversion" is identical to standard field inversion in $\mathbb{F}_{2^8}$ for non-zero field elements with $0^{(-1)} = 0$.

- The intermediate value x is regarded as an $\mathbb{F}_2$ vector of length 8 and transformed using an (8×8) $\mathbb{F}_2$-matrix L_A. The transformed vector $L_A \cdot x$ is then regarded in the natural way as an element of $\mathbb{F}_{2^8}$.

- The output of the AES S-Box is $(L_A \cdot x) + 63$ (here 63 is the usual hexadecimal denotation of the byte 11000011), where addition is with respect to $\mathbb{F}_2$.

The AES linear diffusion (mixing) layer.

- Each row of the array is rotated by a certain number of byte positions.

- Each column of the array is considered to be an $\mathbb{F}_{2^8}$-vector, and a column y is transformed to the column Cy, where C is a (4×4) $\mathbb{F}_{2^8}$-matrix.

In [101] it is shown how to transfer an affine component of an S-Box to the diffusion layer, so that S-Box is represented only by taking the inverse in $\mathbb{F}_{2^8}$. We use this approach in the following discussion.

The AES subkey addition. Each byte of the array is added (with respect to $\mathbb{F}_2$) to the corresponding byte from the corresponding array of round keys.

The round key we have just mentioned are created via the key schedule. Its specification is very similar to the main AES encryption, cf. [102].

In [47] C.Cid *et.al.* proposed the so-called small scaled variants of the AES. The motivation for introducing this notion was that it is impossible to investigate feasibility of algebraic attacks, when one applies them directly to the original AES. So, Cid *et.al.* proposed to scale down the original cipher AES in terms of:

- the number of rounds n ($1 \leq n \leq 10$);

- the number of rows r in the rectangular representation ($r = 1, 2, 4$);

- the number of columns c in the rectangular representation ($c = 1, 2, 4$);

- the size e of a word in bits ($e = 4, 8$).

The notation for the scaled-down cipher is $SR(n, r, c, e)$. In this cipher all rounds are the same which is not quite true for the AES: the $10th$ round differs from the others. But it differs only by an affine mapping, so in principle is the same. Thus we may stick to studying ciphers $SR(n, r, c, e)$. From this prospective, the AES-128 with 10 identical rounds would be the cipher $SR(10, 4, 4, 8)$. In [47] it is written in detail how to scale down the encryption operations, we refer the reader to this paper.

2.4 Attacking the small scale variants of AES via solving a system in key variables only

2.4.1 Analyzing the polynomial system for cryptanalysis

One can write equations for cryptanalyzing (the small scale variants of) the AES directly bitwise over $\mathbb{F}_2$. That was an initial proposal of Courtois and Pieprzyk in [54]. There every byte of a 4×4 array is represented by 8 variables (we have 4 variables for the small scale variants with $e = 4$), each responsible for a corresponding bit in that byte. The equations can then be written in quite a straightforward way. Schematically and abusing notation

2.4. ATTACKING WITH KEY VARIABLES

we can write these equations as

$$w_0 = p + k_0, \tag{2.1}$$
$$SBOX(x_i, w_{i-1}) = 0, \quad i = 1, \ldots, n, \tag{2.2}$$
$$w_i = L(x_i) + k_i, \quad i = 1, \ldots, n, \tag{2.3}$$
$$SBOX_K(s_i, k_{i-1}) = 0, \quad i = 1, \ldots, n, \tag{2.4}$$
$$k_i = L_K(s_i) + L'_K(k_{i-1}) + \kappa_i, \quad i = 1, \ldots, n, \tag{2.5}$$
$$c = L(x_n) + k_n. \tag{2.6}$$

The field equations for all the variables are included. All identifiers are meant to refer to collections of variables (e.g. $w_0 = \{w_{0,0,0}, \ldots, w_{0,0,e-1}, \ldots, w_{0,rc-1,0}, \ldots, w_{0,rc-1,e-1}\}$) except c and p, which are composed of elements in $\mathbb{F}_2$. Here $SBOX, SBOX_K$ are S-Box transformations for the encryption and the key schedule resp.; L, L_K are affine transformations; κ_i are the round constants. Note that operations in (2.2) and (2.4) are done on each separate byte, whereas in (2.1), (2.3), (2.5), and (2.6) are done on the whole rectangular array. The equations in (2.2) and (2.4) are of degree 2. The equations from $SBOX$ arise from $\mathbb{F}_{2^e}$-equations $xw = 1$ translated over $\mathbb{F}_2$ via $x = \sum_{i=0}^{e-1} x_i a^i$ and $w = \sum_{i=0}^{e-1} w_i a^i$, where $m(a) = a^4 + a + 1 = 0$ for the case $e = 4$ and $m(a) = a^8 + a^4 + a^3 + a + 1 = 0$ for the case $e = 8$. So here we suppose that no 0-inversion occurs, which is true with high probability cf. [101]. We now write the S-Box equations are as follows. **S-Box equations for $e = 4$:**

$$x_2w_3 + x_1w_3 + x_3w_2 + x_2w_2 + x_3w_1 + x_0w_0 + 1 = 0,$$
$$x_3w_3 + x_1w_3 + x_2w_2 + x_3w_1 + x_0w_1 + x_1w_0 = 0,$$
$$x_1w_3 + x_2w_2 + x_0w_2 + x_3w_1 + x_1w_1 + x_2w_0 = 0,$$
$$x_1w_3 + x_0w_3 + x_2w_2 + x_1w_2 + x_3w_1 + x_2w_1 + x_3w_0 = 0.$$

S-Box equations for $e = 8$:

$$x_7w_7 + x_6w_7 + x_3w_7 + x_2w_7 + x_1w_7 + x_7w_6 + x_4w_6 + x_3w_6 + x_2w_6 +$$
$$+ x_5w_5 + x_4w_5 + x_3w_5 + x_6w_4 + x_5w_4 + x_4w_4 + x_7w_3 + x_6w_3 + x_5w_3 +$$
$$+ x_7w_2 + x_6w_2 + x_7w_1 + x_0w_0 + 1 = 0,$$
$$x_6w_7 + x_4w_7 + x_1w_7 + x_7w_6 + x_5w_6 + x_2w_6 + x_6w_5 + x_3w_5 + x_7w_4 +$$
$$+ x_4w_4 + x_5w_3 + x_6w_2 + x_7w_1 + x_0w_1 + x_1w_0 = 0,$$
$$x_7w_7 + x_5w_7 + x_2w_7 + x_6w_6 + x_3w_6 + x_7w_5 + x_4w_5 + x_5w_4 + x_6w_3 +$$
$$+ x_7w_2 + x_0w_2 + x_1w_1 + x_2w_0 = 0,$$
$$x_7w_7 + x_2w_7 + x_1w_7 + x_3w_6 + x_2w_6 + x_4w_5 + x_3w_5 + x_5w_4 + x_4w_4 +$$
$$+ x_6w_3 + x_5w_3 + x_0w_3 + x_7w_2 + x_6w_2 + x_1w_2 + x_7w_1 + x_2w_1 + x_3w_0 = 0,$$

$x_7w_7 + x_6w_7 + x_1w_7 + x_7w_6 + x_2w_6 + x_3w_5 + x_4w_4 + x_0w_4 + x_5w_3 +$
$+x_1w_3 + x_6w_2 + x_2w_2 + x_7w_1 + x_3w_1 + x_4w_0 = 0,$
$x_6w_7 + x_3w_7 + x_1w_7 + x_7w_6 + x_4w_6 + x_2w_6 + x_5w_5 + x_3w_5 + x_0w_5 +$
$+x_6w_4 + x_4w_4 + x_1w_4 + x_7w_3 + x_5w_3 + x_2w_3 + x_6w_2 + x_3w_2 + x_7w_1 +$
$+x_4w_1 + x_5w_0 = 0,$
$x_7w_7 + x_4w_7 + x_2w_7 + x_5w_6 + x_3w_6 + x_0w_6 + x_6w_5 + x_4w_5 + x_1w_5 +$
$+x_7w_4 + x_5w_4 + x_2w_4 + x_6w_3 + x_3w_3 + x_7w_2 + x_4w_2 + x_5w_1 + x_6w_0 = 0,$
$x_7w_7 + x_6w_7 + x_5w_7 + x_2w_7 + x_1w_7 + x_0w_7 + x_7w_6 + x_6w_6 + x_3w_6 +$
$+x_2w_6 + x_1w_6 + x_7w_5 + x_4w_5 + x_3w_5 + x_2w_5 + x_5w_4 + x_4w_4 + x_3w_4 +$
$+x_6w_3 + x_5w_3 + x_4w_3 + x_7w_2 + x_6w_2 + x_5w_2 + x_7w_1 + x_6w_1 + x_7w_0 = 0.$

Here by (w_i) and (x_i) we understand the bits coming in and out of an S-Box.

The system presented above and also the BES-system from [101] invoked a lot of research and were a starting point for algebraic cryptanalysis of the AES. Although much effort was put in analyzing the structure of those systems, not much progress is achieved in obtaining competitive attacks. In fact, researchers were only able to cryptanalyze very basic small scale variants of the AES. For example in [44] and [47] the authors could not go further $SR(10,1,1,4), SR(2,1,1,8), SR(4,2,1,4), SR(1,2,2,4)$ for BES equations and $SR(10,1,1,4), SR(2,1,1,8)$ for $\mathbb{F}_2$-equations. Although it can be shown that one can go a bit further, that does not solve our main goal. The XSL method should also be mentioned here [54]. Initially it was believed that this method might be able to give an attack that could, at least theoretically, break AES, but some evidence afterwards showed that estimates behind the XSL method were too optimistic [43].

Everything said above is a motivation for our present work. We need some preparation now. Namely, we will slightly rewrite equations (2.1)-(2.6). In rewriting the equations we aim at the situation where we express every successive variable via its predecessors. So we rewrite equations (2.1)-(2.6) as follows.

$$w_0 = p + k_0, \tag{2.7}$$
$$x_i = sbox(w_{i-1}), \quad i = 1, \ldots, n, \tag{2.8}$$
$$w_i = L(x_i) + k_i, \quad i = 1, \ldots, n, \tag{2.9}$$
$$s_i = sbox_K(k_{i-1}), \quad i = 1, \ldots, n, \tag{2.10}$$
$$k_i = L_K(s_i) + L'_K(k_{i-1}) + \kappa_i, \quad i = 1, \ldots, n, \tag{2.11}$$
$$c = L(x_n) + k_n. \tag{2.12}$$

The field equations on all the variables are added. Here $sbox, sbox_K$ are S-Box transformations for the encryption and the key schedule resp.; L, L_K are affine transformations. How do we achieve the form $x_i = sbox(w_{i-1})$? Recall that initially we have degree-2 equations $SBOX(x_i, w_{i-1}) = 0$. Now we

2.4. ATTACKING WITH KEY VARIABLES

impose a block order, such that $x_i > w_{i-1}$ and find a reduced Gröbner basis for these equations. We obtain exactly $x_i = sbox(w_{i-1})$ and one equation $(w_{i-1,*,0}+1)\ldots(w_{i-1,*,e-1}+1) = 0$, which we can drop out by assuming that the case $w_{i-1,*,0} = 0,\ldots,w_{i-1,*,e-1} = 0$ does not occur (which is true with high probability). The same is true for the key schedule. It is interesting to note that actually one can get rid of equations of the type $(w_{i-1,*,0}+1)\ldots(w_{i-1,*,e-1}+1) = 0$ without any assumptions on non-occurrence of 0-inversion as above. Recall that the S-Box equation $SBOX(x,w) = 0$ is written with an assumption that no 0-inversion occurs. Otherwise, one must rewrite $\mathbb{F}_{2^e}$-equations $x - w^{2^e-2} = 0$ over $\mathbb{F}_2$ to consider also 0-inversions. As we have checked, it turns out that if we write the equations in $SBOX$ this way, we obtain exactly the equations from $sbox$ without $w_{i-1,*,0} = 0,\ldots,w_{i-1,*,e-1} = 0$.

Denote now by G the set of polynomials from (2.7)-(2.11) and the complete set of field equations for every variable. We denote by the same letter an ideal generated by G. Note that G contains both polynomials responsible for encryption/key schedule and the field equations for all the variables. The following holds.

Theorem 2.4.1 *The set G is a Gröbner basis of a zero-dimensional ideal with respect to the lexicographic order induced by $k_0 < w_0 < s_1 < x_1 < k_1 < w_1 < \cdots < s_n < x_n < k_n < w_n$. Variables in each of the variable-blocks $k_0, w_0, \ldots, k_n, w_n$ are ordered arbitrarily.*

Proof. The claim on dimension follows easily form the fact that the field equations for all the variables are included in G. Now note that G consists of the set B of Boolean polynomials with a linear leading term (those coming from encryption/key schedule) and the set F of the field equations. The claim on the Gröbner basis follows by applying the product criterion separately to pairs from the set B and F (Proposition 1.2.12), and then the linear lead criterion to pairs, where one element is from B and another is from F (Proposition 1.2.13). ◇

Remark 2.4.2
- Other elimination orders can be used in Theorem 2.4.1, e.g a block order, where each variable-block constitutes one block.

- Clearly the claim that B is a Gröbner basis is trivial due to the product criterion. A distinguishing feature of Theorem above is that we actually work with the field equations included. So *a priori* it is not clear that G is a Gröbner basis. The result is guaranteed by Proposition 1.2.13. It is crucial for our method, since field equations are always implicitly included in POLYBORI.

Note that when $e = 4$ *sbox* and *sbox$_K$* have degree 3 and when $e = 8$ degree 7. The new equations appear more complex, but the advantage is that we can express each successive variable via its predecessors. The S-Box equations for $e = 4$ follow.

$$x_0 = w_3w_2w_1 + w_2w_1w_0 + w_2w_1 + w_2w_0 + w_3 + w_2 + w_1 + w_0,$$
$$x_1 = w_3w_1w_0 + w_3w_1 + w_2w_1 + w_2w_0 + w_1w_0 + w_3,$$
$$x_2 = w_3w_2w_0 + w_3w_0 + w_2w_0 + w_1w_0 + w_3 + w_2,$$
$$x_3 = w_3w_2w_1 + w_3w_2 + w_3w_1 + w_3w_0 + w_3 + w_2 + w_1.$$

For the case $e = 8$ we have that expressions for x-variables contain resp. 118, 118, 119, 127, 119, 122, 138, 128 monomials.

Different representations for the S-Boxes were studied in the literature (cf. [14, 45]). It is known that up to now "a change of the Rijndael polynomial should not affect the strength of the cipher". It is interesting to see if this is also true in our situation. If we take all 30 irreducible polynomials of degree 8 over $\mathbb{F}_2$, it can be seen that average number of monomials in the S-Box equations for $e = 8$ varies from 122 to 139, which is close to $256/2 = 128$, half of the number of all Boolean monomials of degree up to 7. So we see that changing Rijndael polynomial to some other irreducible polynomial does not essentially decrease the number of terms in the S-Box equations.

2.4.2 Gröbner basis shape. Normal forms

Now let us use the Gröbner-shape of G as per Theorem 2.4.1 to actually obtain equations in initial key variables k_0 only. A quite obvious corollary from Theorem 2.4.1 shows how to do this.

Corollary 2.4.3 *Denote by R the polynomials (equations) in (2.12). For each $f \in R$, $\mathrm{NF}(f, G)$ contains only initial key variables k_0.*

Proof. Indeed, all variables except k_0 appear as leading terms of the Gröbner bases G. Since $r_f := NF(f, G)$ is the reduced normal form, it may not contain any variables except k_0, because $f - r_f$ should be reduced w.r.t G. ◊

Note that we can obtain the same result by simply plugging in the variables successively from the beginning to the end and then plugging the obtained expressions into R. Similar approach was proposed for the BES in [127].

For computing a normal form with respect to a system consisting of polynomials with pairwise different linear leading terms (and the complete set of field equations), there exist fast, highly specialized algorithms in POLYBORI. The Buchberger normal form algorithm ([26]) has the disadvantage that it

2.4. ATTACKING WITH KEY VARIABLES

uses iteration over the leading terms of intermediate results and thus its running time gets a quite direct dependency on the number of terms. The special algorithms in POLYBORIonly depend on the ZDD (zero suppressed decision diagram) structure of the involved polynomials [24]. Basically there are two variants:

1. If we have computed the reduced Gröbner basis, then all polynomials in it are tail reduced, i.e. tails of all the polynomials are reduced w.r.t this Gröbner basis (see Definition 1.2.10), so the computations are faster. But computing the reduced Gröbner basis is quite expensive in this case, since participating polynomials in our case are dense of possibly high degree. Therefore, in our case the tail reductions are expensive.

2. If we work only with a Gröbner basis (not a reduced Gröbner basis), the recursive normal form computation splits in more recursive branches, as the tails of the polynomials (which are used during the computation) still have to be rewritten.

Going back to Corollary 2.4.3 we see that it is possible to eliminate all variables except the initial key variables.

The following timings that show an application of Corollary 2.4.3 have been done on a AMD Dual Opteron 2.2 GHz (we have used only one CPU) with 8 GB RAM on Linux using POLYBORI0.6.1 ([24]). Solving the final system with the key variables is done with the symmgbGF2 algorithm [25] (an advanced version of SLIMGB: [22]) implemented in POLYBORI.

Cipher	t_{red}, sec.	t_{sol}	t_{total}	memory, MB
SR(10,1,1,4)	0.02	0.003	0.023	75
SR(10,1,2,4)	0.29	0.005	0.295	79
SR(10,2,1,4)	0.22	0.005	0.225	87
SR(10,1,1,8)	3.32	0.005	3.325	183
SR(10,2,2,4)	1850	0.01	1850.01	170
SR(2,1,2,8)	68	0.01	68.01	115
SR(3,1,2,8)	8730	0.01	8730.01	145
SR(2,2,1,8)	100	0.2	100.2	116
SR(3,2,1,8)	6550	0.01	6550.01	140

Here t_{red} time needed to do the reductions, t_{sol} time needed to solve the final system in key variables only, t_{total} is the total time of a computation, "memory" shows the memory peaks for the computations. Note that the reduction step takes the lion's share of the computation, whereas time for the final solving is quite negligible. We could not perform the necessary reductions in MAGMA V. 2.15-5 [38]. We tried NormalForm, Reduce,

ReduceGroebnerBasis both with and without field equations. We have also tried the special data structure BooleanPolynomialRing for working with Boolean polynomials. Therewith we could solve only $SR(10, 1, 1, 4)$, for other instances the necessary reductions were infeasible. The reduction was done with respect to lex ordering. PolyBoRI- and Magma-examples together with the running scripts can be downloaded from http://www.mathematik.uni-kl.de/~bulygin/en/files.html.

We tried solving systems with Magma and Singular [74] as per (2.1)-(2.6). We could attack only the full 10 rounds of $SR(10, 1, 1, 4)$ and 3–4 rounds of some 8-bit ciphers. See also results in [47, 44].

We would like to mention an attack based on MRHS linear equations ([109]). In [109] the author was able to break $SR(10, 1, 1, 8)$ in 0.32 sec., which is better than above. Note, however that for key sizes larger than 8, the method of MRHS linear equations needs some bits of a key to be guessed. For instance for $SR(10, 2, 1, 8)$ one needs 8 bits out of 16 to be guessed. Our method does not have such a limitation. It is also notable that for 8 bits known in advance in $SR(10, 2, 1, 8)$ our method needs 4 sec. to execute.

Another approach that is worth mentioning is the recent one from [95]. There the authors also analyze, in particular, the small scaled variants of AES. Using different tricks, the authors were able only to break $SR(10, 2, 1, 4)$ in more than 10000 sec. (0.2 sec. in our table above). Notably, the methods proposed by the authors of [95] experience difficulties already with $SR(3, 2, 2, 4)$, whereas our approach can handle $SR(10, 2, 2, 4)$. Moreover, in [95] only the case $e = 4$ is considered and we manage to obtain also results for the case $e = 8$.

For $SR(10, 2, 2, 4)$ we have 16 equations of degree 16 in 16 key variables with approx. 32000 terms per equation. Note that the equations we obtain are actually reduced with respect to field equations for the key variables. In general with this approach for $SR(10, r, c, e)$, taking an assumption that the small scale variants of AES have the "best possible" diffusion layer, at the end we will have a system of rce equations in rce unknowns of degree rce with the number of terms approx. 2^{rce-1} per equation. So solving such a system for "large" parameters r, c, e is a big challenge. One further point is that PolyBoRIcan represent structured polynomials of huge size in a compact way via the so-called zero suppressed decision diagrams (ZDDs). Not only is this data structure makes it possible to store polynomials in memory efficient way, but also operations on polynomials can be made very fast. More details on that in [24].

Remark 2.4.4 [Using many pairs] Although most of the time we observe that the reductions take the lion's share of the computations, when the num-

2.4. ATTACKING WITH KEY VARIABLES

ber of rounds is small it is not the case. In fact the time needed to do reductions for $SR(2,2,4,4)$ is only 0.15 sec. Nevertheless, it was impossible to solve the resulting system in key variables only in reasonable time. Note that here we may use the fact that the resulting system is in key variables only, so adding more plaintext/ciphertext pairs adds more equations without adding new variables. Since obtaining a set of equations from one pair is fast for $SR(2,2,4,4)$, we pursue this approach. By collecting equations from 8 pairs we were able to break $SR(2,2,4,4)$ in less than 50 sec.: 1.23 sec. totally for reductions and 46.3 sec. for solving. We investigate further this approach in the next section.

2.4.3 Meet-in-the-middle attack

The idea of the meet-in-the-middle attack is not new in cryptanalysis. The ideas of "meet-in-the-middle" were employed already for attacking DES. The main feature of such attacks is to move from both sides: plaintext and ciphertext in the "middle" of a cipher, and there find some binding relations. In algebraic cryptanalysis the technique of "meet-in-the-middle" is studied in [47, 44] in context of the small scale variants of AES. There the authors propose to divide equations for n rounds into two subsystems: one consisting of equations for rounds $1, \ldots, n/2$ (here n is assumed to be even), and the one consisting of equations for rounds $n/2 + 1, \ldots, n$. By computing Gröbner bases with respect to lex ordering the authors get rid of variables that do not appear in rounds $n/2$ and $n/2 + 1$. So at the end one deals with a smaller system with the variables from rounds $n/2$ and $n/2 + 1$ only. Using also equations from the key schedule it is possible to recover the key. We also mention the use of the meet-in-the-middle principle in [1, 2].

In our approach the meet-in-the-middle attack as described in [47, 44] can be realized in quite a straightforward way. One just has to do "usual" reductions in rounds $1, \ldots, n/2$, and then "reverse" reductions in rounds $n/2 + 1, \ldots, n$. One then gets equations in k_0 and k_n from the equations describing encryption plus some key-variables-only equations from the key schedule. In this section our aim will be to illustrate this idea and to see how far can we go in parameters r, c, and e when attacks on two rounds of $SR(2, r, c, e)$ are considered. Note that here we will essentially use equations obtained from several pairs to make our systems more overdetermined. This is possible, since reduction complexity is not an issue here, see below for the results.

Recall the equations (2.7)-(2.12). Let us write them down again and then discuss how to "invert" the second half of the rounds to make the meet-in-the-middle attack possible. For simplicity we assume now that n is even.

Encryption		Key Schedule	
$w_0 = p + k_0$	(E1)	$s_i = sbox_K(k_{i-1})$	(K1,i)
$x_i = sbox(w_{i-1})$	(E2,i)	$k_i = L_K(s_i) + L'_K(k_{i-1}) + \kappa_i$	(K2,i)
$w_i = L(x_i) + k_i$	(E3,i)	$s_j = sbox_K(k_{j-1})$	(K3,j)
$x_j = sbox(w_{j-1})$	(E4,j)	$k_j = L_K(s_j) + L'_K(k_{j-1}) + \kappa_j$	(K4,j)
$w_j = L(x_j) + k_j$	(E5,j)		
$c = L(x_n) + k_n$	(E6)		

Above $i = 1, \ldots, n/2$ and $j = 1+n/2, \ldots, n$. The field equations are assumed to be included. Now doing "usual" reduction in (E1)-(E3,$n/2$) we get equation of the form $w_{n/2} =$ "something in $k_0, \ldots, k_{n/2}$", and in (K1,1)-(K2,$n/2$) of the form $k_{n/2} =$ "something in k_0". Now we need to "invert" equations (E4,j),(E5,j),(E6), (K3,j), and (K4,j) for $j = 1 + n/2, \ldots, n$ to meet these reduced equations in the middle. The former four types of equations are easily invertible, namely (E4,j) inverts as $w_{j-1} = sbox^{-1}(x_j) = sbox(x_j)$, considering that in all $SR(n, r, c, e)$ holds $sbox = sbox^{-1}$. Similarly (K3,j) inverts as $k_{j-1} = sbox_K^{-1}(s_j) = sbox_K(s_j)$. Now (E5,j) inverts as $x_j = L^{-1}(k_j + w_j)$, where L^{-1} is the inverse of the invertible affine transformation L. Similarly (E6) inverts as $x_n = L^{-1}(k_n + c)$.

The question remains only with (K4,j): $k_j = L_K(s_j) + L'_K(k_{j-1}) + \kappa_j$. A standard procedure can be used to do the job. Let us represent this linear system in matrix form:

$$(E_{rce} \mid M \mid N),$$

where E_{rce} is a unity matrix $rce \times rce$, and $M \in Mat(\mathbb{F}_2, rce, re), N \in Mat(\mathbb{F}_2, rce, rce)$ are full-rank matrices. Obviously, E_{rce} corresponds to k_j-variables, M to s_j-, and N to k_{j-1}-variables. Now if we perform Gaussian elimination so that matrix M is brought to the diagonal form, we obtain:

$$\left(\begin{array}{c|c|c} T' & E_{re} & N' \\ \hline T'' & 0 & N'' \end{array} \right).$$

Here $T', N' \in Mat(\mathbb{F}_2, re, rce), T'', N'' \in Mat(\mathbb{F}_2, rce - re, rce)$. Therefore it is possible to get the following linear equations:

$$s_j = A_1^{(j)}(k_{j-1}, k_j), A_2^{(j)}(k_{j-1}, k_j) = 0,$$

where $A_1^{(j)}$ is given by matrices T' and N', and $A_2^{(j)}$ is given by T'' and N''. The block $s_j = A_1^{(j)}(k_{j-1}, k_j)$ has re equations (the number of components

2.4. ATTACKING WITH KEY VARIABLES

in s-variables) and the block $A_2^{(j)}(k_{j-1}, k_j) = 0$ has $rce - re$ equations. So inverting equations in this way we get:

Encryption		Key Schedule	
$w_0 = p + k_0$	(E1)	$s_i = sbox_K(k_{i-1})$	(K1,i)
$x_i = sbox(w_{i-1})$	(E2,i)	$k_i = L_K(s_i) + L'_K(k_{i-1}) + \kappa_i$	(K2,i)
$w_i = L(x_i) + k_i$	(E3,i)	$k_{j-1} = sbox_K(s_j)$	(K3',j)
$w_{j-1} = sbox(x_j)$	(E4',j)	$s_j = A_1^{(j)}(k_{j-1}, k_j)$	(K4',j)
$x_j = L^{-1}(k_j + w_j)$	(E5',j)	$A_2^{(j)}(k_{j-1}, k_j) = 0$	(K4",j)
$x_n = L^{-1}(k_n + c)$	(E6')		

Now obviously equations (E3,$n/2$) and (E4',$1+n/2$) match through w_1-variables which yields rce equations in $k_0, \ldots, k_n$ of total degree $e-1$ from the encryption part. From the key schedule we get rce equations in k_0 and $k_{n/2}$ of total degree $e-1$ from (K2,$n/2$), re equations in k_n and $k_{n/2}$ of total degree $e-1$ from (K3',$1+n/2$) and $(n/2)(rce-re)$ linear equations in $k_{n/2}, \ldots, k_n$ from (K4",j). Now note that if we want to use P plaintext/ciphertext pairs for our attack, then the equations from the key schedule will be the same for all the pairs and the equations from the encryption part will be different. Summarizing we have that using P pairs we get $Prce + rce + re$ equations of degree $e-1$ and $(n/2)(rce-re)$ linear equations. All these equations are composed of $(n+1)rce$ variables, namely coming from $k_0, \ldots, k_n$. It is possible to eliminate (some or all) variables $k_1, \ldots, k_{n/2}$ by applying "mixed" reductions from (E1)-(E3,$n/2$) and (K1,1)-(K2,$n/2$).

In order to show how different the meet-in-the-middle representation is from the one we have in (2.7)-(2.12), let us compare some important parameters of obtained systems, using $SR(2,2,2,4)$ as an example. As it is mentioned above, the system in the initial key variables only from (2.7)-(2.12) obtained via Corollary 2.4.3 includes only k_0 variables, whereas the one constructed using the meet-in-the-middle principle includes k_0, k_1, and k_2. The following table gives the comparison of some parameters (field equations are not considered):

method	# vars	# eqs	highest deg.	av. # of terms
"normal" 1 pair	16	16	9	≈ 2500
"normal" 10 pairs	16	160	9	≈ 2500
"m-i-m" 1 pair	48	48	3	10
"m-i-m" 10 pairs	48	192	3	22

As we can see, although the extent to which the systems are overdetermined is better in the "normal" systems, we gain much in degree and the number of terms by applying the meet-in-the-middle principle.

Let us now present some experimental results that were obtained using POLYBORI. The timings have been done on a AMD Dual Opteron Processor 242 GB 1.6 GHz (we have used only one CPU) with 8 GB RAM on Linux. The same machine has been used for MAGMA experiments.

Cipher	Key, bit	P	t_{red}, sec.	T_{red}, sec.	t_{solve}, sec.	mem., MB
SR(2,2,4,4)	32	8	0.1	0.8	1.3	35
SR(2,4,2,4)	32	8	0.2	1.6	1.4	37
SR(2,4,4,4)	64	8	0.3	2.4	3.7	61
SR(2,2,2,8)	32	64	1.0	130	35	1214
SR(2,1,4,8)	32	64	0.4	25.6	7	353

Here t_{red} is time to obtain key-variables-only equations from one pair via normal form reductions as described above, T_{red} is the total time for reductions of all pairs, and t_{solve} is time to solve the final key-variables-only system. "mem." means the peak of memory consumption. We use fast dense linear algebra implemented in LIBM4RI by Gregory Bard and Martin Albrecht, [94]. Note that results for $SR(2,4,4,4)$ are quite interesting. They imply that we are able to break with only 8 pairs a 64-bit cipher (although a very simple one) in basically 4 sec. if we do the eight reductions in parallel, and 6 sec. without parallelization.

Remark 2.4.5 It is remarkable that meet-in-the-middle principle helps a lot for the examples in the table above. Note that without this principle, i.e. using just approach from Section 2.4.2, we can solve $SR(2,2,2,4)$ with 8 pairs in 13 sec., and $SR(2,2,4,4)$ with 8 pairs in 470 sec. – a remarkable difference. This is justified by the table on page 105.

Let us now present some results obtained with MAGMA 2.14-15.

Cipher	Key, bit	P	t_{red}, sec.	T_{red}, sec.	t_{solve}, sec.	mem., MB
SR(2,2,4,4)	32	8	0.1	0.8	2.5	43
SR(2,4,2,4)	32	8	0.6	4.8	7.8	190
SR(2,4,4,4)	64	8	2.9	23.2	64.8	667

We were unable to get any reasonable results for $SR(2,2,2,8)$ and $SR(2,1,4,8)$. Namely it is quite impossible to make reductions as we need. We tried NormalForm, Reduce, ReduceGroebnerBasis both with and without field equations. Only with
ReduceGroebnerBasis and field equations included could we perform reductions for just one pair's encryption part in 163 seconds.

2.4. ATTACKING WITH KEY VARIABLES

Remark 2.4.6
- Of course it is possible to apply the method for n even rounds to the case of n odd almost with no changes: e.g. one just has to do "forward" reductions in rounds 1–$(n+1)/2$ and "reverse" reductions in the rounds $(n+1)/2+1,\ldots,n$. Specifically we are interested in handling 3 rounds. We could not break even 32-bit ciphers for $e=4$ in any reasonable time with this approach. See the next section on more advanced techniques to tackle the problem.

- Special data structures employed in POLYBORI are particularly useful for normal form reductions that we need for obtaining key-variables-only equations. It seems that general purpose computer algebra systems like MAGMA and SINGULAR are not well suited for the purpose of our approach.

2.4.4 Further optimizations

It is well known (see e.g. [11]), that it is usually easier to compute the Gröbner bases of strongly overdetermined systems. Applying the techniques presented in the previous sections to obtain systems in the key variables only, it is possible to get such systems. The relation between variables and equations can be improved by considering more pairs of ciphertext and plaintext. However the number of variables is still quite large.

The equation systems obtained in Section 2.4.3 seem to have more structure than those in Section 2.4.2 (at least for a small number of rounds). However most Gröbner bases implementations (even if they are optimized for the Boolean case) are quite unaware of this structure. Assuming, that we have a subsystem, which involves much less variables (which is for certain a result of the weak diffusion layer when working with small number of rounds like 2 or 3), the usual Buchberger's algorithm would mix them quite strongly with the polynomials that are not included in the considered subsystem, so that their combination would involve more variables. Since more variables usually make computations harder, such an effect is not desirable. So it seems quite appropriate to treat these subsystems in less variables separately to preserve their structure. However, they could describe a complex zero set, whose knowledge might not solve our problem (usually we assume that the complete system has exactly one solution). Actually it is possible in our experiments to overdetermine these subsystems by increasing the number of pairs, so that they will also have only one solution in practice. In this way, solving them will yield the solution for every variable involved in the subsystem. Using these values, it will be much easier to solve the complete system by computing a Gröbner basis or applying the same trick again. Of course clustering subsystems of polynomials does not seem to be obvious in general.

Finding a good (not necessarily the best) solution for identifying such subsystems in polynomial time seems to be related to optimization techniques. In our case, we just used a very simple solution: we considered the variable sets of each of the Boolean polynomials that constitute the system. Let $f_i, i = 1, \ldots, N$ for some N be non-linear polynomials in the key variables only system. For each $i = 1, \ldots, N$ we examine which variables occur in f_i. Denote this set of variables by $Var(f_i)$. Then we collect all the polynomials from the system that contain only variables from $Var(f_i)$ and no others. Denote the subsystem so obtained by $SubSys(f_i)$. Then among all $SubSys(f_i), i = 1, \ldots, N$ we look only for overdetermined ones. Among these overdetermined ones we collect those with the least number of variables occurring, i.e. with the smallest $|Var(f_i)|$. Then for example we can choose the one subsystem that has among those the most polynomials, i.e. the one with maximal $|SubSys(f_i)|$. Combining this subsystem with the linear equations in key variables only we can solve for $Var(f_i)$. This makes the clustering very fast. Of course, the performance of our computations strongly depends on finding a subsystem in as few variables as possible and also as much overdetermined as possible. While the practical experiments show that the initial idea is quite promising, our method here is quite basic and should be optimized for more complex ciphers to give better results (easier equation systems) to be processed by our solving algorithms.

Below we present some results of obtaining overdetermined subsystems. Although exact values for the number of variable and equations in an overdetermined subsystems depends on the given plaintext/ciphertext pairs, the following table provides a realistic picture for the situation.

Cipher	P	N	# vars	min $|Var(f_i)|$	max $|SubSys(f_i)|$
SR(3,2,4,4)	64	8232	128	33	256
SR(2,2,2,8)	64	2096	96	33	65
SR(2,1,4,8)	64	2088	96	17	64
SR(2,2,4,8)	256	16464	192	33	256
SR(2,4,4,4)	64	4176	192	33	64

Note that in order to observe such an effect, having many plaintext/ciphertext pairs is important. For instance, for the cipher $SR(2, 4, 4, 4)$ with 64 pairs we could detect several such overdetermined subsystems, whereas with only 8 pairs there were none.

Remark 2.4.7 • We operate here with the term "overdetermined system". By this we mean, of course, not only that just the number of

2.4. ATTACKING WITH KEY VARIABLES

equations is larger, than the number of variables, but that the above property holds for the maximal number of linearly independent equations. In particular, while trying to obtain subsystems of smaller size, sometimes we encountered a situation when the "rank of the system" was slightly smaller than the number of equations. In general, though, rank and the number of equations do coincide.

- As we may see from the table above, the ratio "#eqs/#vars" drops as we go from the initial system to a smaller overdetermined system. Nevertheless, it seems that it makes sense to choose the system as small as possible (even with a smaller ratio), since solving is then significantly faster. Still sometimes a more careful inspection of subsystems may be necessary: it may happen (and is actually observed) that a slightly larger subsystem yields a significantly larger ratio of overdetermination.

Another aspect often applied in cryptanalysis is the guessing of bits in the key. This can be combined quite well with the idea introduced in the previous subsection. Having found a subsystem in less number of variables, we consider small subsets of these variable (preferably some variables out of the initial key k_0). Plugging in trial values for these variables can yield a system, which is much easier to solve via Gröbner basis computations for each possible value of a trial. This means, that in the presented timings, we really try each combination of these small sets of bits we guess, so no knowledge of them is required for the attack in advance. For example for $SR(2,2,2,8)$ and 64 pairs plugging in just two bits simplified the systems dramatically, so that it was easier to solve four of these systems, than just one initial system. While plugging in values provides some speed up, Gröbner basis computations can be seen as a good supplement to these search techniques. This is the area, where still much research needs to be done, in a much wider scope: touching both the field of computational algebra as well as the very optimized toolset of SAT-solvers. An initial (quite theoretical) publication on these aspects is [48].

We do not have a special selection strategy for the bits at the moment. We tried just taking the first bits of the k_0 variables in the subsystem, as well as a random selection. For the instances we considered, taking random bits sometimes turned out to be not very productive. Finding a more sophisticated method here could yield further improvements of results. The only thing, we can say at this point of time, that it is seems reasonable to guess bits of k_i only for one i. We next present some results obtained with the methods of this section.

Cipher	Key, bit	P	t_{red}, sec.	T_{red}, sec.	t_{solve}, sec.	# b.g.
SR(3,2,4,4)	32	256	0.01	2.56	121	7
SR(2,2,4,8)	64	256	2.2	1227	236	10

Here "# b.g." means the number of bits guessed during the computations.

Let us see how the computation goes for the cipher $SR(3,2,4,4)$ with 256 plaintext/ciphertext pairs and 7 bits as a guessing parameter. After doing all necessary reductions for each from the 256 pairs, the smaller overdetermined subsystem is identified. As we have seen, it is possible to find an overdetermined subsystem with 33 variables and 256 equations. The variables of this subsystem are $k_{0,4-7}, k_{0,16-19}, k_{0,24-31}, k_{1,16-19}, k_{1,28-31}, k_{2,19}, k_{0,16-23}$, where $k_{i,j}$ means j-th key bit at the round i. We then do an exhaustive search on the first 7 bits from the initial key k_0 in the subsystem. So consider all $2^7 = 128$ combinations, where the variables $k_{0,25-31}$ are assigned all possible values. Therewith we have to solve 128 times: first the subsystem, and then if it has a Gröbner basis not equal to $\{1\}$, pass the solution(s) to help to solve the large system. Since we have 256 pairs, it is not a surprise that only for one assignment out of 128 do we get a solution for the subsystem. This solution yields 33 values of key variables out of 128. Thus final solving is now feasible, which is reflected in the table.

Conclusions and future work

This chapter was devoted to algebraic cryptanalysis of the small scale variants of the AES. We concentrated on the study of the systems composed of the key variables only. This facilitates, among other things, the use of several plaintext/ciphertext pairs. We saw an effect of such use on the example of attacking 2-3 rounds via the meet-in-the-middle strategy. Here the following directions seem to be of interest.

- Further adjusting of data structures in POLYBORIto the needs of algebraic cryptanalysis.

- Combination of algebraic ideas with the ideas of conventional cryptanalysis as is done, e.g. in [1, 2].

- Algebraic cryptanalysis of stream ciphers proved to be a rather effective tool. Looking closely at stream ciphers, especially non-LFSR ciphers appears to be interesting.

- Applying algebraic techniques in analyzing other cryptographic primitives, such as hash functions, seems to be promising.

Bibliography

[1] M. Albrecht, C. Cid, "Algebraic Techniques in Differential Cryptanalysis", available at http://eprint.iacr.org/2008/177.pdf, 2008.

[2] M. Albrecht, C. Cid, "Algebraic Techniques in Differential Cryptanalysis", *Proceedings of the First International Conference on Symbolic Computation and Cryptography, Beijing, China*, pp.55–60, 2008.

[3] S. Arimoto, "Encoding and decoding of p-ary group codes and the correction system," (in Japanese) *Inform. Processing in Japan*, vol. 2, pp. 320–325, Nov. 1961.

[4] D. Augot, "Description of minimum weight codewords of cyclic codes by algebraic system," *Finite Fields Appl.*, vol. 2, no.2, pp.138–152, 1996.

[5] D. Augot, P. Charpin, N. Sendrier, "The minimum distance of some binary codes via the Newton's Identities," *Eurocodes'90, LNCS 514*, pp. 65–73, 1990.

[6] D. Augot, P. Charpin and N. Sendrier, "Studying the locator polynomial of minimum weight codewords of BCH codes," *IEEE Trans. Inform. Theory*, vol. IT-38, pp. 960–973, May 1992.

[7] D. Augot, M. Bardet, J.-C. Faugère, "Efficient Decoding of (binary) Cyclic codes beyond the correction capacity of the code using Gröbner bases," *INRIA Report*, no. 4652, Nov. 2002.

[8] D. Augot, M. Bardet, J.C. Faugère, "On formulas for decoding binary cyclic codes", *Proc. IEEE Int. Symp. Information Theory*, 2007.

[9] D. Augot, M. Bardet, J.C. Faugère, "On the decoding of cyclic codes with Newton identities", to appear in *Special Issue "Gröbner Bases Techniques in Cryptography and Coding Theory" of Journ. Symbolic Comp.*, 2008.

[10] M. Bardet, J.-C. Faugère, B. Salvy, "Complexity of Gröbner basis computation for semi-regular overdetermined sequences over $GF(2)$ with solutions in $GF(2)$", *INRIA Report*, no. 5049, 2003.

[11] M. Bardet and J.-C. Faugère and B. Salvy, "On the complexity of Gröbner basis computations of semi-regular overdetermined algebraic equations," *ICPSS'2004*.

[12] M. Bardet, J.-C. Faugère, B. Salvy, B-Y. Yang "Asymptotic behaviour of the index of regularity of quadratic semi-regular polynomial systems," *in MEGA 2005, Eighth International Symposium on Effective Methods in Algebraic Geometry, Porto Conte, Alghero, Sardinia (Italy), May 27th - June 1st,* 2005.

[13] A. Barg, "Complexity issues in coding theory," *in "Handbook on Coding Theory", V.S Pless and W.C. Huffmann eds.*, pp.649–754, 1998.

[14] E. Barkan, E. Biham, "In How Many Ways Can You Write Rijndael?", *ASIACRYPT 2002*, LNCS vol.2501, pp.160–175, 2002.

[15] E.R. Berlekamp, "Algebraic coding theory," *Mc Graw Hill*, New York, 1968.

[16] E.R. Berlekamp, R.J. McEliece, H.C.A. van Tilborg, "On the Inherent Intractability of Certain Coding Problems, " *IEEE Transactions on Information*, Vol.IT-24, no. 3, pp. 384–386, May 1978.

[17] M.A. de Boer, R. Pellikaan, "Gröbner bases for codes," in *Some tapas of computer algebra* (A.M. Cohen, H. Cuypers and H. Sterk eds.), Chap. 10, pp. 237–259, Springer-Verlag, Berlin 1999.

[18] M.A. de Boer, R. Pellikaan, "Gröbner bases for decoding," in *Some tapas of computer algebra* (A.M. Cohen, H. Cuypers and H. Sterk eds.), Chap. 11, pp. 260–275, Springer-Verlag, Berlin 1999.

[19] M. Borges-Quintana, M. A. Borges-Trenard, P. Fitzpatrick, E. Martínez-Moro, "Gröbner bases and combinatorics for binary codes," *Applicable Algebra in Engineering, Communication and Computing*, vol. 19, no. 5, pp. 393–411, 2008.

[20] M. Borges-Quintana, M. A. Borges-Trenard, E. Martínez-Moro, "A General Framework for Applying FGLM Techniques to Linear Codes," *Lecture Notes in Computer Science*, vol. 3857, pp. 76–86, 2006.

[21] P. Bours, J.C.M. Janssen, M. van Asperdt, J.C.A. van Tilborg, "Algebraic decoding beyond e_{BCH} of some binary cyclic codes when $e > e_{BCH}$," *IEEE Trans. Inform. Theory*, vol. IT-36, pp. 214–222, Jan. 1990.

[22] M. Brickenstein, "Slimgb: Gröbner Bases with Slim Polynomials", *Zentrum für Computeralgebra, Kaiserslautern*, September, 2005.

[23] M. Brickenstein, S. Bulygin, "Attacking AES via Solving Systems in the Key Variables Only." *Proceedings of the First International Conference on Symbolic Computation and Cryptography, Beijing, China, April 28-30*, pp. 118–123, 2008.

[24] M. Brickenstein, A. Dreyer, "PolyBoRi: A framework for Gröbner basis computation with Boolean polynomials", to appear in *Special Issue "MEGA'2007" of Journal of Symbolic Computation*, 2008.

[25] M. Brickenstein, A. Dreyer, G. Greuel, M. Wedler, O. Wienand "New developments in the theory of Gröbner bases and applications to formal verification" *Preprint*, available at http://arxiv.org/abs/0801.1177, 2008.

BIBLIOGRAPHY

[26] B. Buchberger, "Ein Algorithmus zum Auffinden der Basiselemente des Restklassenrings nach einem nulldimensionalen Polynomideal", *Universität Innsbruck*, Dissertation, 1965.

[27] J. Buchmann, "Introduction to Cryptography", *Springer*, 2004.

[28] J. Buchmann, A. Pyshkin, R.-P. Weinmann, "A zero-dimensional Groebner basis for AES-128", *FSE 2006, LNCS 4047*, pp. 78–88, 2006.

[29] J. Buchmann, A. Pyshkin, R.-P. Weinmann, "Block Ciphers Sensitive to Groebner Basis Attacks", *CT-RSA 2006, LNCS 3860*, pp. 313–331, Springer-Verlag, 2006.

[30] S. Bulygin. decodegb.lib. A SINGULAR 3.1 library for Decoding and min distance of linear codes with GB (2008).

[31] S. Bulygin, R. Pellikaan, "Bounded distance decoding of linear error-correcting codes with Gröbner bases", to appear in *Journal of Symbolic Computation Special Issue "Gröbner Bases Techniques in Cryptography and Coding Theory"*, 2009.

[32] S. Bulygin, R. Pellikaan, "Decoding error-correcting codes with Groebner bases". *Proceedings of the 28-th Symposium on Information Theory in the Benelux, Enschede, The Netherlands, May 24-25*, pp. 3–10, 2007.

[33] S. Bulygin, R. Pellikaan, "Decoding linear error-correcting codes up to half the minimum distance with Gröbner bases", to appear as a short note in *"Gröbner Bases, Coding, and Cryptography"*, RISC Book Series, Springer, 2009.

[34] S. Bulygin, R. Pellikaan, "Decoding and finding the minimum distance with Gröbner bases: history and new insights", to appear as a chapter in *I. Woungang, S. Misra, S.C. Misra (Eds.) "Selected Topics in Information and Coding Theory"*, World Scientific, 2009.

[35] S. Bulygin, M. Brickenstein, "Obtaining and solving systems of equations in key variables only for the small variants of AES", submitted to *Mathematics in Computer Science Special Issue "Symbolic Computation and Cryptography"*, available at http://eprint.iacr.org/2008/435, 2009.

[36] W. Bruns, U. Vetter, "Determinantal rings," *Lect. Notes in Math.*, vol. 1327, Springer-Verlag, Berlin 1988.

[37] M. Caboara, T.Mora, "The Chen-Reed-Helleseth-Truong decoding algorithm and the Gianni-Kalkbrenner Gröbner shape theorem," *Appl. Algeb. Eng. Commum. Comput.*, 13, pp.209–232, 2002.

[38] J. J. Cannon, W. Bosma (Eds.), "Handbook of Magma Functions", Edition 2.14 (2007).

[39] X. Chen, I.S. Reed, T. Helleseth, T.K. Truong, "Algebraic decoding of cyclic codes: a polynomial point of view," *Contemporary Math.* vol. 168, pp. 15–22, 1994.

[40] X. Chen, I.S. Reed, T. Helleseth, T.K. Truong, "Use of Gröbner bases to decode binary cyclic codes up to the true minimum distance," *IEEE Trans. Inform. Theory*, vol. IT-40, pp. 1654–1661, Sept. 1994.

[41] X. Chen, I.S. Reed, T. Helleseth, T.K. Truong, "General principles for the algebraic decoding of cyclic codes," *IEEE Trans. Inform. Theory*, vol. IT-40, pp. 1661–1663, Sept. 1994.

[42] C. Cid (Ed.), "Algebraic Cryptanalysis of Symmetric Primitives", *Security report of the European Network of Excellence in Cryptology*, 2008.

[43] C. Cid, G. Leurent, "An Analysis of the XSL Algorithm", *In B.Roy, editor, Advances in Cryptology - ASIACRYPT 2005*, vol. 3788 of LNCS, pp.333–352, 2005.

[44] C. Cid, S. Murphy, M.J.B. Robshaw, "Algebraic Aspects of the Advanced Encryption Standard", *Springer-Verlag*, 2006.

[45] C. Cid, S. Murphy, M. Robshaw, "An Algebraic Framework for Cipher Embeddings", *Proceedings of the 10th IMA International Conference on Coding and Cryptography*, LNCS 3796, pp. 278–289, 2005.

[46] C. Cid, S. Murphy, M. Robshaw, "Computational and Algebraic Aspects of the Advanced Encryption Standard", *Seventh International Workshop on Computer Algebra in Scientific Computing, CASC 2004*, pp. 93–103, St. Petersburg, Russia, 2004.

[47] C. Cid, S. Murphy, M. Robshaw, "Small Scale Variants of the AES", *Fast Software Encryption - FSE2005, LNCS 3557*, pp. 145–162, 2005.

[48] M. Clegg, J. Edmonds, and R. Impagliazzo, "Using the Groebner basis algorithm to find proofs of unsatisfiability", *Proceedings of the Twenty-eighth Annual ACM Symposium on the Theory of Computing*, pp. 174-183, 1996.

[49] CoCoATeam, "CoCoA: a system for doing Computations in Commutative Algebra", Available at http://cocoa.dima.unige.it.

[50] J.T. Coffey, R.M. Goodman, P.G. Farrell, "New approaches to reduced-complexity decoding", *Discrete Applied Mathematics*, no.33, pp.43–60, 1991.

[51] A.B. Cooper, "Toward a new method of decoding algebraic codes using Gröbner bases," *Trans. 10th Army Conf. Appl. Math. and Comp.,*, pp.1–11, 1993.

[52] , N. Courtois, A. Klimov, J. Patarin, A. Shamir, "Efficient algorithms for solving overdefined systems of multivariate polynomial equations", *Lecture Notes in Computer Science*, vol. 1807, pp. 392–407, 2000.

[53] , N. Courtois, J. Patarin, "About the XL algorithm over GF(2)", *Lecture Notes in Computer Science*, vol. 2612, pp. 141–157, 2003.

[54] N. Courtois, J. Pieprzyk, "Cryptanalysis of Block Ciphers with Overdefined Systems of Equations", in *Asiacrypt 2002, LNCS 2501*, pp. 267–287, Springer, 2002.

[55] D. Cox, J. Little, D. O'Shea, "Ideals, varieties, and algorithms", 2nd Edition, *Springer-Verlag*, 1997.

[56] J. Daemen, R. Vincent, "The Design of Rijndael", *Springer*, 2002.

[57] S.M. Dodunekov, J.E.M. Nilsson, "Algebraic decoding of Zetterberg codes", *IEEE Trans. Inform. Theory*, vol. IT-38, pp. 1570–1573, Sept. 1992.

[58] J.L. Dornstetter, "On the equivalence of Berlekamp's and Euclid's algorithm," *IEEE Trans. Inform. Theory*, vol. IT-33, pp. 428–431, May 1987.

[59] D. Eisenbud, "Linear sections of determinantal varieties," *Amer. J. Math.*, vol. 110, pp. 541–575, 1988.

[60] D. Eisenbud, "Commutative algebra with a view toward a algebraic geometry," *Grad. Texts in Math. vol. 150, Springer-Verlag, New York*, 1995.

[61] J. Farr, S. Gao, "Gröbner bases, Padé approximation, and decoding of linear codes," *Coding Theory and Quantum Computing*, Contemporary Mathematics, vol. 381, pp. 3–18, 2003.

[62] J. Farr, S. Gao, "Gröbner bases and generalized Padé approximation," *Mathematics of Computation*, vol. 75, pp. 461–473, 2005.

[63] J.C.Faugère, "A new efficient algorithm for computing Gröbner bases (F4)," *Journal of Pure and Applied Algebra*, 139(1–3), pp.61–88, 1999.

[64] J.C.Faugère, "A new efficient algorithm for computing Gröbner bases without reduction to zero F5," In T. Mora, editor, *Proceedings of the 2002 International Symposium on Symbolic and Algebraic Computation ISSAC*, pp. 75–83, 2002.

[65] J.-C. Faugère, P.Gianni, D. Lazard, T.Mora, "Efficient Computation of Zero-dimensional Gröbner Bases by Change of Ordering", *J. Symb. Comput.*, 16, pp.329–344, 1993.

[66] FGb, http://fgbrs.lip6.fr/jcf/Software/FGb/index.html.

[67] P. Fitzpatrick, "On the key equation", *IEEE Transactions on Information Theory*, 41, no. 5, pp.1290–1302, 1995.

[68] P.Fitzpatrick, J.Flynn, "A Gröbner basis technique for Padé approximation," *J.Symbolic Computation*. 24(5), pp.133–138, 1992.

[69] J. Fitzgerald, "Applications of Gröbner bases to Linear Codes," *Ph.D. Thesis*, Louisiana State University, 1996.

[70] J. Fitzgerald and R.F. Lax, "Decoding affine variety codes using Gröbner bases," *Designs, Codes and Cryptography*, vol. 13, pp. 147–158, 1998.

[71] GAP – Groups, Algorithms, and Programming, Version 4.4, The GAP Group, www.gap-system.org, 2006.

[72] M. Giorgetti, M. Sala, "A commutative algebra approach to linear codes", *BCRI preprint no.58*, www.bcri.ucc.ie, 2006.

[73] D.C. Gorenstein and N. Zierler, "A class of error-correcting codes in p^m symbols," *Journ. SIAM*, vol. 9, pp. 207–214, 1961.

[74] G.-M. Greuel, G. Pfister, and H. Schönemann. SINGULAR 3.0 — A computer algebra system for polynomial computations. In M. Kerber and M. Kohlhase: *Symbolic computation and automated reasoning, The Calculemus-2000 Symposium* (2001), pages 227–233.

[75] G.-M. Greuel, G. Pfister, "A SINGULAR Introduction to Commutative Algebra", 2nd Edition, *Springer Verlag*, 2008.

[76] C.R.P. Hartmann and K.K. Tzeng, "Decoding beyond the BCH bound using multiple sets of syndrome sequences," *IEEE Trans. Inform. Theory*, vol. IT-20, pp. 292–295, Mar. 1974.

[77] A.E. Heydtmann, J.M. Jensen, "On the equivalence of the Berlekamp-Massey and the Euclidean algorithms for decoding," *IEEE Trans. Inform. Theory*, vol. 46, pp. 2614–2624, Nov. 2000.

[78] R.J. Higgs, J.F. Humphreys, "Decoding of the ternary Golay code," *IEEE Trans. Inform. Theory*, vol. IT-39, pp. 1043–1046, May 1992.

[79] T. Høholdt, J.H. van Lint, R. Pellikaan, "Algebraic geometry codes," in *Handbook of Coding Theory*, vol. 1, pp. 871–961, (V.S. Pless and W.C. Huffman eds.), Elsevier, Amsterdam 1998.

[80] J.A. van der Horst and T. Berger, "Complete decoding of triple-error-correcting binary BCH codes," *IEEE Trans. Inform. Theory*, vol. IT-22, pp. 138-147, Mar. 1976.

[81] J.F. Humphreys, "Algebraic decoding of the ternary [13,7,5] quadratic residue code," *IEEE Trans. Inform. Theory*, vol. IT-38, pp. 1122–1125, May 1992.

[82] D. Joyner, "GUAVA: A GAP4 Package for computing with error-correcting codes," Version 3.1, 2007, *http://www.gap-system.org/Manuals/pkg/guava3.1/htm/chap0.html*

[83] P. Källquist, "Decoding Zetterberg codes," *Proc. Fourth Joint Swedisch-Soviet Workshop on Inform. Theory*, Gotland, Sweden, pp. 305–309, 1989.

[84] A.I. Kostrikin and I.R. Shafarevich (Eds.), "Algebra I: Basic Notions of Algebra", *Encyclopedia of Mathematical Sciences*, vol. 11, *Springer-Verlag*, 1990.

[85] E. Kunz, *Introduction to commutative algebra and algebraic geometry*, Birkhauser, Boston 1985.

[86] S. Lang, "Algebra", 3rd Edition, *Addison-Wesley Publishing*, 1993.

[87] R. Lidl, H. Niederreiter "Introduction to finite fields and their applications", *Cambridge University Press*, 2000.

BIBLIOGRAPHY

[88] J.H. van Lint, "Introduction to Coding Theory", *Springer-Verlag*, 2nd Edition, 1992.

[89] J. Little, "A key equation and the computation of error values for codes from order domains," http://arxiv.org/abs/math/0303299, 2003.

[90] C.Lossen, A.Frühbis-Krüger, "Introduction to Computer Algebra (Solving Systems of Polynomial Equations)" http://www.mathematik.uni-kl.de/~lossen/SKRIPTEN/COMPALG/compalg.ps.gz, 2005.

[91] P. Loustaunau, E.V. York, "On the decoding of cyclic codes using Gröbner bases," *AAECC*, vol. 8 (6), pp. 469–483, 1997.

[92] M.G. Marinari, H.M. Möller, T. Mora, "Gröbner basis of ideals defined by functional with an application to ideals of projective points," *Appl. Algebra Engrg. Comm. Comput.*, 4, no. 2, pp. 103–145, 1993.

[93] F.J.MacWilliams, N.J.A. Sloane, "The Theory of Error-correcting Codes", *Amsterdam-New York-Oxford: North Holland*, 1977.

[94] Martin Albrecht and Gregory Bard, The M4RI Team "The M4RI Library – Version 20080624", Website http://m4ri.sagemath.org, 2008.

[95] C. V. Miolane and L. R. Knudsen, "Block Cipher Analysis" *Academic dissertation, in series: (ISBN:)*, pages: 176, 200902.

[96] J.L. Massey, "Shift-register synthesis and BCH decoding," *IEEE Trans. Inform. Theory* vol. IT-15, pp. 122–127, Jan. 1969.

[97] R.J. McEliece, "A public-key cryptosystem based on algebraic coding theory," *DSN Progress Report*, vol. 42-44, pp. 114–116, 1978.

[98] A. Menezes, P. van Oorschot, S. Vanstone, "Handbook of Applied Cryptography", *CRC Press*, available online http://www.cacr.math.uwaterloo.ca/hac/, 1996.

[99] T. Mora, E. Orsini, "Decoding cyclic codes: the Cooper philosophy", *a talk at the Special Semester on Gröbner bases*, May, 2006.

[100] T. Mora, M.Sala, "On the Groebner bases for some symmetric systems and their application to coding theory," *J. Symb. Comp.*, vol.35, no.2, p.177–194, 2003.

[101] S. Murphy, M. Robshaw, "Essential Algebraic Structure within the AES", *Advances in Cryptology – CRYPTO 2002, Lecture Notes in Computer Science 2442, M. Yung Ed. Springer, Berlin*, pp. 1–16, 2002.

[102] National Institute of Standards and Technology. Advanced Encryption Standard. FIPS 197. 26 November 2001.

[103] H. Niederreiter, "Knapsack-type crypto systems and algebraic coding theory," *Problems of Control and Information Theory*, vol. 15 (2), pp. 159–166, 1986.

[104] R. Overbeck, N. Sendrier, "Code-based cryptography," in *D.J. Bernstein, J. Buchmann, E. Dahmen (Eds.)"Post-Quantum Cryptography,"Springer*, pp. 95–145, 2009.

[105] R. Pellikaan, "On the existence of error-correcting pairs," *Journal of Statistical Planning and Inference*, vol. 51, pp. 229–242, 1996.

[106] R. Pellikaan, X.-W. Wu, S. Bulygin, "Codes and Cryptography on Algebraic Curves", a book in progress, to be published by *Cambridge University Press*, 2010.

[107] W.W. Peterson, "Encoding and error-correction procedures for the Bose-Chauduri codes," *IRE Trans. Inform. Theory*, vol. IT-6, pp. 459–470, 1960.

[108] W.W. Peterson, E.J. Weldon, "Error-correcting codes," *MIT Pres, Cambridge*, 1977.

[109] H. Raddum, "MRHS Equation Systems", in *LNCS 4876*, pp. 232–245, 2007.

[110] I.S. Reed, X. Yin, T.K. Truong, "Algebraic decoding of the (32,16,8) quadratic residue code," *IEEE Trans. Inform. Theory*, vol. IT-36, pp. 876–880, July 1990.

[111] I.S. Reed, T.K. Truong, X. Chen, X. Yin, "Algebraic decoding of the (41,21,9) quadratic residue code," *IEEE Trans. Inform. Theory*, vol. IT-38, pp. 974–986, May 1992.

[112] T.G. Room, "The geometry of determinantal loci", *Cambridge University Press, Cambridge*, 1938.

[113] E. Orsini, M.Sala, "Correcting errors and erasures via the syndrome variety," *J. Pure and Appl. Algebra*, 200, pp.191–226, 2005.

[114] E.Orsini, M.Sala, "General error locator polynomials for binary cyclic codes with t$\leq$2 and n<63," *IEEE Trans. Inform. Theory*, vol. 53, no. 3, pp. 1095–1107 2007.

[115] E. Orsini, M. Sala, "Improved decoding of affine–variety codes", *BCRI preprint no.68*, www.bcri.ucc.ie, 2007.

[116] M.Sala, "Gröbner basis techniques to compute weight distributions of shortened cyclic codes," *J. Algebra Appl.*, vol. 6, no. 3, pp. 403–414, 2007.

[117] P.W. Show, "Polynomial-time algorithms for prime factorization and discrete logarithms on a quantum computer," *SIAM Journal on Computing*, 26(5), pp. 1484-1509, 1997.

[118] I.E. Shparlinski, "Finding irreducible and primitive polynomials," *Appl. Alg. Engin. Commun. Comp.* vol. 4, pp. 263–268, 1993.

[119] I.E. Shparlinski, "Finite fields: Theory and computation," *Mathematics and its Applications*, vol. 477, Kluwer Acad. Publ., Dordrecht, 1999.

[120] A.Shamir, J.Patarin, N.Cortois, A.Klimov, "Efficient algorithms for solving overdetermined systems of multivariate polynomial equations," *Advances in cryptology - EUROCRYPT'00*, vol. 1807 of Lecture notes in Computer Science, pp.392–407, 2000.

[121] C.E. Shannon, "Communication Theory of Secrecy Systems", *Bell System Technical Journal*, 28-4:656–715, 1949.

BIBLIOGRAPHY

[122] V.M. Sidelnikov, S.O. Shestakov, "On the insecurity of cryptosystems based on generalized Reed-Solomon codes," *Discrete Math. Appl.*, vol. 2, pp. 439–444, 1992.

[123] Singular Manual, available at
http://www.singular.uni-kl.de/Manual/latest/index.htm

[124] P. Stevens, "Extensions of the BCH decoding algorithm to decode binary cyclic codes up to their maximum error correction capacities," *IEEE Trans. Inform. Theory*, vol. IT-34, pp. 1332–1340, 1988.

[125] D.R. Stinson, "Cryptography Theory and Practice", *CRC Press*, 1995.

[126] Y. Sugiyama, M. Kasahara, S. Hirasawa and T. Namekawa, "A method for solving the key equation for decoding Goppa codes," *Information and Control*, vol. 27, pp. 87–99, 1975.

[127] I. Toli, A. Zanoni, "An Algebraic Interpretation of AES-128", in *Advanced Encryption Standard AES: 4th International Conference, AES 2004, Revised Selected and Invited Papers. Hans Dobbertin, Vincent Rijmen, Aleksandra Sowa editors.* LNCS 3373, pp. 84–97, http://dx.doi.org/10.1007/11506447_8.

[128] K.K. Tzeng, C.R.P. Hartmann, R.T. Chien, "Some notes on iterative decoding," *Proc. 9th Allerton Conf. Circuit and Systems Theory*, Oct. 1971.

[129] A.Vardy, "The intractability of computing the Minimum Distance of a Code", *IEEE Trans. Inform. Theory*, Vol. IT-43, no. 6, pp. 1757–1766, November 1997.

[130] B.-Y. Yang, J.-M. Chen, "All in the XL Family: Theory and Practice," in *C. Park and S. Chee (Eds.): ICISC 2004, LNCS 3506*, pp. 67–86, 2005.

BIBLIOGRAPHY

Index

adaptive chosen-ciphertext, 93
attack
    adaptive chosen-ciphertext, 93
    adaptive chosen-plaintext, 93
    chosen-ciphertext, 93
    chosen-plaintext, 93
    ciphertext-only, 93
    known-plaintext, 93
    related-key, 93

cipher
    block, 91
        iterative, 92
    Caesar, 91
    permutation, 92
    stream, 92
    substitution, 91
    transposition, 92
ciphertext, 91
code
    affine variety, 27
    cyclic, 7
        complete defining set, 7
        defining set, 7
    dimension, 4
    equivalent, 4
    information rate, 4
    length, 4
    linear, 4
    orthogonal, 5
    perfect, 6
    redundancy, 4
codeword, 4
    nearest, 8

complexity coefficient, 69
confusion, 92
Cooper's philosophy, 14, 17
coset leader, 8
CRHT
    ideal, 17
    syndrome variety, 17
cryptosystem
    symmetric, 91

decoding, 4
    bounded up to half the minimum distance, 8
    covering sets, 8
    formal, 24
    generic, 15, 24
    information sets, 8
    maximum-likelihood, 6
    one-step, 24
    online, 15, 25
decryption, 91
diffusion, 92

encoding, 4
encryption, 91
error vector, 7
error-correcting capacity, 6
error-locator polynomial, 22
    general, 18
        multi-dimensional, 29
exhaustive search, 8

generalized Newton identities, 22
generalized power sum function, 16

Gröbner basis, 11
    reduced, 11

key, 91
    decryption, 91
    encryption, 91
    schedule, 92
    secret, 91

leading
    coefficient, 11
    ideal, 11
    monomial, 11
    term, 11
linearization, 69
    extended, 70

matrix
    generator, 4
    Macaulay, 65
    parity check, 5
minimum
    distance, 5
        relative, 5
    weight, 5
monomial order, 10
    block, 10
    degree reverse lexicographic, 10
    elimination, 11
    lexicographic, 10
    product, 10

plaintext, 91
polynomial
    Rijndael, 95

round transformation, 92

S-Box, 92
standard
    monomial, 11
syndrome, 8, 16
    decoding, 8

known, 16
polynomial, 23
unknown, 16

I want morebooks!

Buy your books fast and straightforward online - at one of the world's fastest growing online book stores! Environmentally sound due to Print-on-Demand technologies.

Buy your books online at
www.get-morebooks.com

Kaufen Sie Ihre Bücher schnell und unkompliziert online – auf einer der am schnellsten wachsenden Buchhandelsplattformen weltweit!
Dank Print-On-Demand umwelt- und ressourcenschonend produziert.

Bücher schneller online kaufen
www.morebooks.de

OmniScriptum Marketing DEU GmbH
Heinrich-Böcking-Str. 6-8
D - 66121 Saarbrücken
Telefax: +49 681 93 81 567-9

info@omniscriptum.com
www.omniscriptum.com

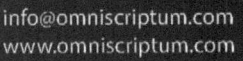

Printed by Books on Demand GmbH, Norderstedt / Germany